Latency and Distortion of Electromagnetic Trackers for Augmented Reality Systems

Synthesis Lectures on Algorithms and Software in Engineering

Editor
Andreas Spanias, *Arizona State University*

Latency and Distortion of Electromagnetic Trackers for Augmented Reality Systems
Henry Himberg and Yuichi Motai
2014

Bandwidth Extension of Speech Using Perceptual Criteria
Visar Berisha, Steven Sandoval, and Julie Liss
2013

Control Grid Motion Estimation for Efficient Application of Optical Flow
Christine M. Zwart and David H. Frakes
2013

Sparse Representations for Radar with MATLAB® Examples
Peter Knee
2012

Analysis of the MPEG-1 Layer III (MP3) Algorithm Using MATLAB
Jayaraman J. Thiagarajan and Andreas Spanias
2011

Theory and Applications of Gaussian Quadrature Methods
Narayan Kovvali
2011

Algorithms and Software for Predictive and Perceptual Modeling of Speech
Venkatraman Atti
2011

Adaptive High-Resolution Sensor Waveform Design for Tracking
Ioannis Kyriakides, Darryl Morrell, and Antonia Papandreou-Suppappola
2010

MATLAB® Software for the Code Excited Linear Prediction Algorithm: The Federal Standard-1016
Karthikeyan N. Ramamurthy and Andreas S. Spanias
2010

OFDM Systems for Wireless Communications
Adarsh B. Narasimhamurthy, Mahesh K. Banavar, and Cihan Tepedelenliouglu
2010

Advances in Modern Blind Signal Separation Algorithms: Theory and Applications
Kostas Kokkinakis and Philipos C. Loizou
2010

Advances in Waveform-Agile Sensing for Tracking
Sandeep Prasad Sira, Antonia Papandreou-Suppappola, and Darryl Morrell
2008

Despeckle Filtering Algorithms and Software for Ultrasound Imaging
Christos P. Loizou and Constantinos S. Pattichis
2008

Latency and Distortion of Electromagnetic Trackers for Augmented Reality Systems
Henry Himberg and Yuichi Motai

ISBN: 978-3-031-00394-3   paperback
ISBN: 978-3-031-01522-9   ebook

DOI 10.1007/978-3-031-01522-9

A Publication in the Springer series
SYNTHESIS LECTURES ON ALGORITHMS AND SOFTWARE IN ENGINEERING

Lecture #12
Series Editor: Andreas Spanias, *Arizona State University*
Series ISSN
Print 1938-1727   Electronic 1938-1735

Latency and Distortion of Electromagnetic Trackers for Augmented Reality Systems

Henry Himberg
Polhemus, Inc., Vermont

Yuichi Motai
Virginia Commonwealth University

SYNTHESIS LECTURES ON ALGORITHMS AND SOFTWARE IN ENGINEERING #12

ABSTRACT

Augmented reality (AR) systems are often used to superimpose virtual objects or information on a scene to improve situational awareness. Delays in the display system or inaccurate registration of objects destroy the sense of immersion a user experiences when using AR systems. AC electromagnetic trackers are ideal for these applications when combined with head orientation prediction to compensate for display system delays. Unfortunately, these trackers do not perform well in environments that contain conductive or ferrous materials due to magnetic field distortion without expensive calibration techniques. In our work we focus on both the prediction and distortion compensation aspects of this application, developing a "small footprint" predictive filter for display lag compensation and a simplified calibration system for AC magnetic trackers.

In the first phase of our study we presented a novel method of tracking angular head velocity from quaternion orientation using an Extended Kalman Filter in both single model (DQEKF) and multiple model (MMDQ) implementations. In the second phase of our work we have developed a new method of mapping the magnetic field generated by the tracker without high precision measurement equipment. This method uses simple fixtures with multiple sensors in a rigid geometry to collect magnetic field data in the tracking volume. We have developed a new algorithm to process the collected data and generate a map of the magnetic field distortion that can be used to compensate distorted measurement data.

KEYWORDS

delta quaternion, extended Kalman filter, head orientation, head tracking, interacting multiple model estimator, quaternion prediction, AC magnetic tracking, magnetic tracker, calibration, position, orientation, real-time measurement, distributed body sensor, motion tracking, human-computer interaction

Contents

List of Tables

Preface

Virtual reality and augmented reality (VR/AR) environments can be used to improve spatial information and situational context from limited sensory data through data fusion and visual presentation, immersing the operator in the simulation environment. Augmented reality systems are often used with Head Mounted Displays (HMDs) in military applications where identification, status and targeting information is superimposed on the real world to improve situational awareness and decrease response time in hostile situations. HMDs are also used to train technicians for a broad section of tasks ranging from welding to ultrasound imaging.

Display lag in simulation environments with helmet-mounted displays causes a loss of immersion that degrades the value of virtual/augmented reality training simulators. Simulators use predictive tracking to compensate for display lag, preparing display updates based on anticipated head motion. These applications require head trackers that can support high measurement rates in the range of 120 Hz to 240 Hz with good precision and a tolerance for a moving frame of reference (some simulators are motion controlled). AC electromagnetic trackers are well suited for head tracking but are affected by conductive and ferromagnetic materials. To be successful in these applications, a prediction capability must be added to the magnetic tracker and a method of compensating for distortion of the magnetic field developed.

Adding a predictive capability to an AC magnetic tracker is not a trivial matter; these devices have limited computational resources and operate with strict real-time constraints. A new method for predicting head orientation based in quaternion orientation was developed for this application and demonstrated to provide the desired prediction capability in production units. The predictor operates on the change in quaternion between consecutive data frames (the delta quaternion), avoiding the heavy computational burden of the quaternion motion equation. Head velocity is estimated from the delta quaternion by an extended Kalman filter and then used to predict future head orientation. Experimental results indicate that the delta quaternion method provides the accuracy of the quaternion method without the heavy computational burden. This predictor has been implemented in the Polhemus Liberty tracker and is currently being demonstrated for several prospective customers.

The first predictor developed as part of this project is able to deal with most head motion but has difficulty providing accurate prediction during aggressive head motion. The reduced computational requirement of the delta quaternion method provides an opportunity to improve prediction performance with multiple model filtering. A new head orientation prediction technique was developed based on delta quaternion filters in a multiple model framework to track angular head velocity and angular head acceleration. The multiple model filter tracks head velocity more closely than the single DQ and additionally, generates angular acceleration estimates that are used in a new prediction algorithm. The prediction algorithm combines the output of the multiple filters using a weighting scheme based on the mode probabilities of the filter and predicts future quaternion orientation.

In addition to dealing with the prediction requirement of the target application, the effect of conductive or ferrous materials in the tracking volume must be dealt with. Careful control of the tracking volume and correct positioning of the transmitter/receiver modules can minimize distortion but in many applications significant distortion remains. Tracking performance can be improved by correcting the position and orientation (PnO) measurements with a compensation factor generated from a mapping procedure. Mapping a tracker volume is a protracted process using large fixtures and requiring hundreds of high precision measurements to assemble a map of the distorted magnetic field. A new field mapping which removes field distortion as part of the PnO computation was developed to simplify the mapping process. This method uses two fixtures with multiple sensors in a rigid geometry to measure the field distortion at a given pose, and simultaneously solves the PnO of all sensors. Collected data is processed off-line to create a look-up table (LUT) for use with various compensation schemes.

Henry Himberg and Yuichi Motai
May 2014

Acknowledgments

The authors would like to thank Polhemus, Inc. (Colchester, Vermont) for the support, time, and equipment used in this study. We would also like to thank Herb Jones and James Farr of Polhemus for their guidance on issues dealing with mapping in tracking environments. Ben Himberg provided great assistance with several 3-D analysis problems in the development of the IVC. This study was supported in part by the School of Engineering at Virginia Commonwealth University and National Science Foundation.

Henry Himberg and Yuichi Motai
May 2014

CHAPTER 1

Delta Quaternion Extended Kalman Filter

1.1 INTRODUCTION

Head tracking is widely used in augmented and virtual reality simulation environments (AR/VR) to control scene rendering in response to head orientation. The perceived latency (lag) between head motion and display response causes a loss of immersion for the user that can result in dizziness in extreme cases [3, 8–10, 14, 15, 22–24]. In training applications, the user learns to compensate for the display latency of the particular simulator, adjusting head motion to improve performance. This learned behavior compensates for display latency in simulation environment but differences between the simulator latency and that of the actual system reduce the value of the training. An effective method of compensating for simulation latency in helmet-mounted Display (HMD) simulators is to predict the future orientation of the head. If head orientation can be accurately predicted, the simulator can render the next scene before the user moves. Various prediction methods have been proposed for latency compensation [1–5, 7, 12, 28, 30, 33] with the Kalman filter receiving considerable attention.

A new method of head orientation prediction, the Delta Quaternion (DQ) Framework, was developed for latency compensation. The DQ framework predicts future head orientation from the change in quaternion orientation between measurements (the delta quaternion). Angular head velocity is estimated from the delta quaternion by an EKF and then combined with the current quaternion measurement to predict future orientation. The DQ differs from other head orientation prediction methods in several ways, including estimation of the delta quaternion instead of the quaternion orientation in the EKF, and decoupling of the prediction interval from the input data rate. Removing the quaternion orientation from the Kalman filter reduces the number of state variables from 7 to 3 in a filter that uses the constant velocity motion model, providing a significant savings of computational resources. The decoupled prediction algorithm avoids a reduction in frame rate required to accommodate the one-step prediction method used in other approaches. Predictive filtering, including the Extended Kalman Filter (EKF), Particle Filters (PF) and the Unscented Kalman Filter (UKF) are widely used for latency compensation [1, 2, 5, 7, 25, 32]. The UKF requires additional computation resource without improving performance when compared to the EKF in estimating quaternion motion [26]. The PF does not provide a significant improvement upon the EKF when used for head motion prediction [7].

The EKF in used in this study of head motion prediction to avoid the additional computational burden of other methods [26, 27].

1.2 RELATED WORK

The author previously developed two adaptive EKF methods for prediction of quaternion head motion in a simulation environment [11]. The first method used a fading memory algorithm to modify the EKF predicted error covariance in response to changes in the filter residual. The algorithm improved tracking performance but increased output noise in some conditions. A second method (R-Adaptive) adaptively modified the measurement covariance to control the output noise level. The R-Adaptive approach provided lowered output noise and improved tracking performance with benign data but had increased RMS error with aggressive head motion.

Kiruluta, Eizenman and Pasupathy proposed a system that used a Kalman filter to predict head motion from position data [1]. The study compared a constant acceleration Kalman filter predictor to a polynomial approach. Experimental results showed that the Kalman filter provided good latency compensation for moderate motion but had degraded performance undergoing fast motion. An adaptive version of the Kalman predictor was also studied for applications requiring tracking of fast motion at the cost of throughput delay.

Goddard [4] and Bohg [3] both proposed methods of orientation prediction based on the quaternion motion equation presented by Chou [16]. Each of these methods predicted future orientation as a function of angular head velocity and current head orientation. The large state vector of the quaternion filter (7 state variables) and the non-linear state equation led to large matrices in the EKF, resulting in a large computational load on the host system.

A head tracking system was developed by Chang and Cho to control camera movement in a surveillance system application [5]. The proposed system used image-based head tracking to track an individual in a defined physical space. A Kalman filter was used to improve stability by predicting head position in the image space.

Liang, Shaw and Green developed a quaternion method of head motion prediction based on Kalman filtering [28]. They based their work on the assumption that perceived latency was mainly caused by the delay in orientation data. The proposed system predicted head orientation using a linearization of the quaternion orientation to break the quaternion into four independent components. Each of the four decoupled components was predicted using a separate Kalman filter. The four predicted components were combined to form a predicted unit quaternion value.

Azuma and Bishop developed a predictive tracking system for a Helmet Mounted Display (HMD) using inertial sensors mounted to the display with Kalman filtering [30]. The system improved latency in most conditions, as compared to prediction without the inertial sensors or no prediction at all.

A comparison of a Grey theory-based prediction algorithm, a Kalman filter approach and an extrapolation method was performed by Wu and Ouhyoung [33]. They found that both the Grey theory method and the Kalman filter significantly improved performance as compared to extrapolation. The authors stated that the Grey theory method performed equally well as the Kalman filter while having a relatively low computation complexity. The computational demands were not qualitatively compared in the study.

Laviola proposed a latency compensation method based on double exponential smoothing as an alternative to Kalman filter prediction [34]. The proposed algorithm was compared to derivative-free Kalman filters (systems without a velocity or acceleration measurements) and found to provide similar performance with a reduced computation requirement.

A phase lead filter system was proposed by So and Griffin to compensate for delays in helmet-mounted displays [2]. The study found that phase lead filters significantly improved head tracking performance but introduced jitter under some conditions. An additional compensation technique using image deflection was used to compensate for filter jitter.

Zhang and Zhou used an adaptive Kalman filter for human movement tracking in medical rehabilitation [12]. The proposed system uses a Kalman filter to control a camera that captures body movement.

Quaternion estimation is often used for attitude control in spacecraft. Ali et al. used a system based on delta quaternions to control attitude in the Mars Exploration Rover [37]. The system applies a heading adjustment to the previous attitude to estimate current orientation. Using the new estimate, the system conducts a series of confirmation tests to determine if the attitude estimate is correct. This system uses a variety of sensors including accelerometers, gyroscopes, wheel odometry and visual odometry to determine vehicle orientation. Similar to our proposal, this system estimates the change in orientation (delta quaternion) and then corrects it based on measurement data. Cheon and Kim used an Unscented Kalman Filter (UKF) to estimate spacecraft attitude with quaternions [21]. This study successfully used magnetometer and gyroscopic data to estimate quaternion orientation with a UKF.

Marins et al. [35] developed an orientation sensor based on a MARG (Magnetic, Angular Rate and Gravity) sensor using Kalman filtering. The study proposed two methods of determining position and orientation from MARG measurement data using Kalman filters. Another study conducted by Sabatini proposed the use of a similar sensor (gyroscope, accelerometer and magnetometer) to measure orientation [36].

Attitude control systems develop and control orientation using Kalman filter with a variety of measurements techniques including gyroscopes, magnetometers and accelerometers. These applications are based on the same quaternion motion equations that we use in our work but differ greatly in the application specifics. Orientation measurement devices use angular rate data to estimate orientation using a Kalman filter, although the specifics of the filter design vary considerably with sensor type. These applications differ from our study in that we are using quaternions with a Kalman filter to estimate angular rate information from an orientation measurement. Our approach is independent of the sensor type used for the measurement. Although we have performed our experiment using the Polhemus tracker, any other method of measuring orientation could be used without loss of performance (assuming similar measurement accuracy).

1.3 BACKGROUND ON ORIENTATION PREDICTION

1.3.1 EXTENDED KALMAN FILTER

The Extended Kalman Filter (EKF) is a prediction-correction filter used in systems with a state Equation (1.1) measurement Equation (1.2).

$$x\left(k|k-1\right)=f\left(x\left(k-1\right),u\left(k-1\right),w\left(k\right)\right) \tag{1.1}$$
$$z\left(k\right)=h\left(x\left(k|k-1\right),v\left(k\right)\right) \tag{1.2}$$

The state Equation (1.1) expresses the state x at time k as a function of the state at time $k-1$, an external input u and process noise w (process noise is defined as any change in state not modeled by the state equation). The measurement Equation (1.2) relates the measurement z at time k to the state at time k and measurement noise v. The process noise and measurement noise are assumed to be independent Gaussian random variables with zero mean [6, 13, 20, 31].

The EKF equations can be applied to non-linear systems using a Taylor expansion to increase the linearity of the system about the current state (1.3) (1.4). The A, W, H and V Jacobian matrices are recomputed each time the filter iterates.

$$x\left(k\right)\approx x\left(k|k-1\right)+A\left(k\right)\cdot\left(x\left(k-1\right)-x\left(k|k-1\right)\right)+W\left(k\right)\cdot w\left(k-1\right) \tag{1.3}$$
$$z\left(k\right)\approx z\left(k|k-1\right)+H\left(k\right)\cdot\left(x\left(k-1\right)-x\left(k|k-1\right)\right)+V\left(k\right)\cdot v\left(k-1\right) \tag{1.4}$$

1.3.2 QUATERNIONS

Unit quaternions are a commonly used method of orientation representation that avoids the singularities of Euler angles and the stability problems of direction cosine matrices [17–19, 25, 29]. A unit quaternion is a four-dimensional representation of orientation that characterizes an orientation as a rotation θ about an axis of rotation defined by the unit vector u (1.5).

$$q=\left[\ \cos\left(\tfrac{\theta}{2}\right)\quad u\cdot\sin\left(\tfrac{\theta}{2}\right)\ \right]^{T} \tag{1.5}$$

Quaternions provide a compact, efficient method of conducting 3D rotations. To rotate an object, the orientation $q(k)$ of the object is multiplied by the desired change in rotation, the delta quaternion Δq (1.6).

$$q\left(k\right)=\Delta q\cdot q\left(k-1\right) \tag{1.6}$$

1.4 FILTER DESIGN

The Q and DQ frameworks estimate angular head velocity ω from measured quaternion orientation q, predicting future orientation as a function of the estimated head velocity and a user-specified prediction time δ. The Quaternion (Q) Framework uses an EKF to estimate current head velocity ω_k from quaternion measurement data $q(k)$. Future orientation $q(k\tau + \delta)$ is predicted as a function of the current quaternion measurement $q(k)$, the corrected angular velocity $\omega(k|k)$, the frame time τ and the prediction interval δ. The Q framework is a two-step process that estimates future orientation directly from quaternion orientation measurement data using a Kalman filter and a prediction function (fP) (Fig. 1.1). The Delta Quaternion (DQ) framework uses a similar process that operates on the delta quaternion between measurements (Fig. 1.2). The DQ framework first converts the incoming data to delta quaternions (Δq) which are then used by an EKF to estimate angular head velocity ω. The DQ framework uses the same prediction function as the Q, calculating the delta quaternion Δq of the prediction interval and applying it to the current quaternion measurement $q(k)$.

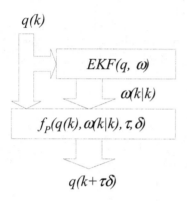

Figure 1.1: The Q framework is a two-step process that estimates future orientation directly from quaternion orientation measurement data using a Kalman filter and a prediction function (fP).

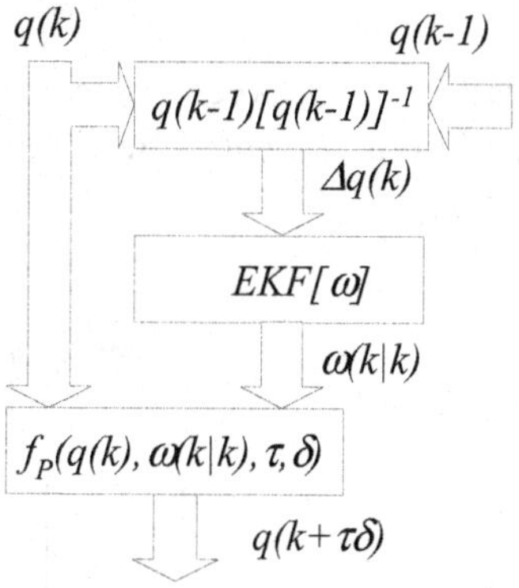

Figure 1.2: The DQ framework is a three-step process that converts quaternion orientation measurements into delta quaternions. Future orientation is predicted using an EKF and a prediction function (f_P).

1.4.1 MOTION MODEL

A constant velocity motion model (CV) is used for each of the frameworks that were investigated. Both the DQ and Q frameworks are based on the change in quaternion being a function of angular velocity. At the high data rate of an AC magnetic tracker, the CV model is a good choice for slow to moderate head motion. The CV model assumes that angular velocity ω is constant from frame to frame using a white noise acceleration component w and the frame period τ to handle any changes in velocity the may occur (1.7).

$$\omega(k) = \omega(k-1) + w \cdot \tau \tag{1.7}$$

1.4.2   QUATERNION FRAMEWORK

The Quaternion Kalman filter uses a state vector consisting of the corrected quaternion orientation $\hat{q}$ ($q(k|k)$) and the corrected angular velocity vector ($\omega(k|k)$) (1.8). The state Equation (1.9) predicts the next state from the current one using the CV model. The measurement Equation (1.10) is linear since the quaternion orientation is included in the state vector.

$$x(k|k-1) = [q(k|k-1)\omega(k|k-1)]^T \tag{1.8}$$

$$f_g(x, w, \tau) = \begin{bmatrix} q(x, w, \tau) \\ \omega + w \cdot \tau \end{bmatrix}^T \tag{1.9}$$

$$h(x, v) = q(k|k-1) + v(k) \tag{1.10}$$

The relationship between quaternion motion and angular velocity using quaternion multiplication (1.11) was presented by Chou [16] where Ψ is the 4×4 element, angular velocity quaternion (1.12).

$$\dot{q} = \Psi + q \tag{1.11}$$

$$\Psi = \frac{1}{2} \cdot \begin{bmatrix} 0 & -\omega_0 & -\omega_1 & -\omega_2 \\ \omega_0 & 0 & -\omega_2 & -\omega_1 \\ \omega_1 & \omega_2 & 0 & -\omega_0 \\ \omega_2 & \omega_1 & \omega_0 & 0 \end{bmatrix} \tag{1.12}$$

As shown in Goddard [4], the solution to this differential equation is an exponential function (1.13) that can be solved for the closed discrete form by assuming constant velocity (1.14).

$$q(t + \tau) = e^{\Psi \cdot \tau} \cdot q(t) \tag{1.13}$$

$$q(k) = \Delta q(\omega(k), \tau) \cdot q(k-1) \tag{1.14}$$

The discrete form rotates the current orientation (q) by a delta quaternion Δq which is a function of angular velocity ω and time τ. The delta quaternion is computed in its compact four-element column vector form (1.15) and expanded to a 4×4 matrix for multiplication operations.

$$\Delta q(\omega, \tau) = \begin{bmatrix} \cos\left(\frac{\theta}{2}\right) & \frac{2 \cdot \omega}{\|\omega\|} \cdot \sin\left(\frac{\theta}{2}\right) \end{bmatrix} \tag{1.15}$$

$$\theta = \tau \cdot \sqrt{\omega^T \cdot \omega}$$

$$\theta = \tau \cdot \sqrt{\omega^T \cdot \omega}$$

The predicted angular velocity at time step k ($\omega(k|k-1)$) is generated with the CV motion model as a function of the corrected angular velocity state from the previous time step ($\omega(k-1|k-1)$), process noise $w(k)$ and the frame period τ (1.16).

$$\omega\,(k|k-1) = \omega\,(k-1|k-1) + w \cdot \tau \tag{1.16}$$

The predicted quaternion state ($q(k|k-1)$) is calculated as the product of the delta quaternion (Δq) generated from the predicted angular velocity ($\omega(k|k-1)$) and the corrected quaternion from time step $k-1$ state ($q(k-1|k-1)$) (1.17).

$$q\,(k|k-1) = \Delta q\,(\omega\,(k|k-1)\,,\tau) \cdot q\,(k-1|k-1) \tag{1.17}$$

The Kalman filter requires four Jacobian matrices (A, W, H and V) to be computed each time the filter iterates. The A matrix contains the partial derivative of the predicted state ($q(k|k-1)$) with respect to each corrected variable from the previous time step ($q(k-1|k-1)$ and $\omega(k-1|k-1)$) and requires three non-trivial partial derivatives (1.18).

$$A_{i,j}\,(k) = \left[\frac{\partial}{\partial x_j}[f\,(x\,(k|k-1)\,,w\,(k))]_i \right]_{w=0}$$

$$A_{i,j}\,(k) = \left[\begin{array}{cc} \frac{\partial}{\partial q_j} q_i\,(k|k-1) & \frac{\partial}{\partial \omega_j} q_i\,(k|k-1) \\ 0 & \frac{\partial}{\partial \omega_j} \omega_i\,(k|k-1) \end{array} \right]_{w=0} \tag{1.18}$$

The partial derivative of the predicted quaternion with respect to the corrected quaternion ($q(k|k)$) is the predicted delta quaternion ($\Delta q(k|k-1)$) (1.19). In this instance the delta quaternion must be expanded to its full 4×4 matrix format for inclusion in the A matrix.

$$\frac{\partial}{\partial q_j} q_i\,(k|k-1) = \frac{\partial}{\partial q_j}[\Delta q\,(\omega\,(k|k-1)\,,\tau) \cdot q\,(k-1|k-1)]_i$$

$$\frac{\partial}{\partial q_j} q_i\,(k|k-1) = \Delta q\,(\omega\,(k|k-1)\,,\tau) \tag{1.19}$$

The partial derivative of the predicted quaternion ($q(k|k-1)$) with respect to the corrected velocity from the previous time step ($w(k-1|k-1)$) is calculated as three column vectors (1.20).

$$\frac{\partial}{\partial \omega} q\,(k|k-1) = \left[\begin{array}{ccc} \frac{\partial}{\partial \omega_0} q\,(k|k-1) & \frac{\partial}{\partial \omega_1} q\,(k|k-1) & \frac{\partial}{\partial \omega_2} q\,(k|k-1) \end{array} \right] \tag{1.20}$$

Starting with the definition of the predicted quaternion ($q(k|k-1)$) (1.15), each four-element column vector is the product of the partial derivative of the predicted delta quaternion with respect to the correct velocity ($\omega(k-1|k-1)$) and corrected quaternion ($q(k|k-1)$) (1.21).

$$\frac{\partial}{\partial \omega_i} q\,(k|k-1) = \left(\frac{\partial}{\partial \omega_i} \Delta q\,(\omega\,(k|k-1)\,,\tau)\right) \cdot q\,(k-1|k-1) \qquad (1.21)$$

A generalized form of the partial derivative of the predicted delta quaternion ($\Delta q\,(k|k-1)$) with respect to velocity state ($\omega(k-1|k-1)$) can be expressed as a function of the predicted delta quaternion, the predicted angular velocity ($w(k|k-1)$) and the time interval τ (1.22).

$$\frac{\partial}{\partial \omega_j} [\Delta q\,(\omega\,(k|k-1)\,,\tau)] = \begin{bmatrix} -\frac{\tau}{4} \cdot \Delta q'_{i+1} \\ \frac{\omega'_0}{\Omega^2} \cdot \left(\tau \cdot \omega(k|k-1)_j \cdot \Delta q'_0 - \Delta q'_{i+1}\right) + (\delta_{i,0}) \cdot \frac{1}{\omega'_i} \cdot \Delta q'_{i+1} \\ \frac{\omega'_1}{\Omega^2} \cdot \left(\tau \cdot \omega'_i \cdot \Delta q'_0 - \Delta q'_{i+1}\right) + (\delta_{i,1}) \cdot \frac{1}{\omega'_i} \cdot \Delta q'_{i+1} \\ \frac{\omega'_2}{\Omega^2} \cdot \left(\tau \cdot \omega'_i \cdot \Delta q'_0 - \Delta q'_{i+1}\right) + (\delta_{i,2}) \cdot \frac{1}{\omega'_i} \cdot \Delta q'_{i+1} \end{bmatrix}$$
$$(1.22)$$

$$\omega' = \omega\,(k|k-1) \quad \Delta q' = \Delta q\,(\omega\,(k|k-1)\,,\tau)$$

$$\delta_{i,j} = \{dirac\}\,(i,j) \quad \Omega = \sqrt{\omega(k|k-1)^T \cdot \omega\,(k|k-1)}$$

Finally, the partial derivative of predicted angular velocity with respect to the velocity state can be obtained by inspection (1.23).

$$\frac{\partial}{\partial \omega_i} \omega' = I \qquad (1.23)$$

The matrix W is the Jacobian of partial derivatives of the predicted state ($q\,(k|k-1)$) with respect to the process noise ($w(k)$) (1.24). The CV motion model simplifies W considerably since the predicted velocity ($\omega(k|k-1)$) is a linear function of the velocity state ($\omega(k-1|k-1)$) and the process noise ($w(k)$). Closer inspection indicates it is the product of (1.21) and the time step τ; resulting in a simplified form (1.25).

$$W_{i,j}\,(k) = \left[\frac{\partial}{\partial w_j} f_i\,(x\,(k|k-1)\,,w\,(k))\right]_{w=0}$$

$$W_{i,j}\,(k) = \left[\frac{\partial}{\partial w_j} \begin{bmatrix} q_i\,(k|k-1) \\ \omega_i\,(kl|k-1) \end{bmatrix}\right]_{w=0} \qquad (1.24)$$

$$W = \begin{bmatrix} \left[\frac{\partial}{\partial \omega} q\,(\omega\,(k|k-1))\right] \cdot \tau \\ \tau \cdot I \end{bmatrix} \qquad (1.25)$$

The H matrix is the Jacobian of partial derivatives of the measurement Equation (1.10) with respect to state ($x\,(k|k)$) which can be derived by inspection (1.26).

$$H_{i,j}(k) = \left[\begin{array}{cc} \frac{\partial}{\partial q_j} h_i\left(x\left(k|k-1\right), v[k]\right) & \frac{\partial}{\partial \omega_j} h_i\left(x\left(k|k-1\right), v(k)\right) \end{array} \right]_{v=0}$$

$$H = \left[\begin{array}{cc} I & 0 \end{array} \right] \tag{1.26}$$

The V matrix is the Jacobian of partial derivatives of the measurement Equation (1.10) with respect to measurement noise. Since the measurement model is linear, V is a 4×4 identity matrix (1.27).

$$V_{i,j}(k) = \left[\frac{\partial}{\partial v_j} h_i\left(x\left(k|k-1\right), v(k)\right) \right]_{v=0}$$

$$V_{i,j}(k) = I \tag{1.27}$$

The Q framework uses a seven-element state vector to estimate angular velocity. Close examination of the quaternion filter equation reveals that the delta quaternion Δq is the driving equation of the filter. All information concerning the change in orientation is contained in the delta quaternion, with the quaternion state providing a method or converting the delta quaternion to match the measurement.

1.4.3   DELTA QUATERNION FRAMEWORK

The DQ framework removes the quaternion equation from the estimation process by directly converting incoming quaternion data $(q(k))$ to the delta quaternion $(\Delta q(k))$ before using the EKF. The EKF now predicts the angular head velocity $(\omega(k))$ directly from the delta quaternion $(\Delta q(k))$. The quaternion motion Equation (1.14) is only needed to compute the predicted quaternion $q(k\tau + \delta)$ and is moved outside the Kalman filter into the orientation prediction process (Fig. 1.3).

The Delta Quaternion (DQ) framework estimates angular head velocity directly from a delta quaternion measurement without estimation of the quaternion itself. Eliminating the quaternion orientation from the Kalman filter reduces the state vector from 7 elements to 3 when using the CV motion model. The resulting reduction in matrix rank (from 7×7 to 3×3) results in a large savings of computational resources while retaining the quaternion motion model (1.14).

The delta quaternion of the current frame $\Delta q(k))$ represents the change in quaternion between the previous frame at time $k-1$ and the current frame at time k. The delta quaternion is computed as the quaternion product of the current quaternion and the inverse of the previous quaternion (1.28).

$$\Delta q(k) = q(k) \cdot \left(q(k-1)\right)^{-1} \tag{1.28}$$

The DQ Kalman filter uses a three-element state vector (1.29) containing the average angular velocity.

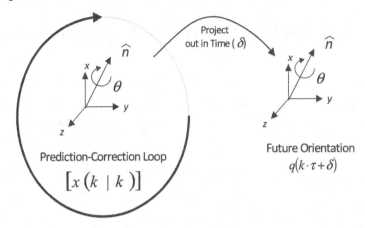

Figure 1.3: The prediction-correction loop of the Kalman filter provides an estimate of angular head velocity which is projected across the prediction interval to estimate the change in orientation (Δq) the will occur.

$$x\,(k) = [\omega\,(k)] \tag{1.29}$$

The CV state equation is now a linear function of the corrected angular velocity from the previous time step ($\omega(k-1|k-1)$), the process noise ($w(k)$) and the time interval τ (1.30).

$$f_{DQ}\,(x\,(k)\,,w\,(k)) = \omega\,(k) + w\,(k)\cdot\tau \tag{1.30}$$

The measurement model in the DQ Kalman filter must relate the predicted angular head velocity ($\omega(l|k-1)$) to the delta quaternion measurement ($\Delta q(k)$). The equation used for the delta quaternion prediction in the quaternion EKF (1.15) is used as the measurement equation for the DQ EKF (1.31).

$$h\,(x,v) = \Delta q\,(\omega,\tau) + v$$
$$h\,(x,v) = \left[\;\cos\left(\tfrac{\theta}{2}\right)\quad \tfrac{2\cdot\omega}{\|\omega\|}\cdot\sin\left(\tfrac{\theta}{2}\right)\;\right] + v \tag{1.31}$$

It should be noted that both the DQ and Q frameworks compute the difference between the measured and predicted quaternion as a simple subtraction which is technically not a valid quaternion operation. The small time interval between input data samples minimizes the effect of this compromise.

Due to the linear state equation, the DQ A and W matrices are constant and do not have to be computed for each iteration of the Kalman filter (1.32) (1.33).

$$A_{i,j}(k) = \frac{\partial}{\partial \omega_j}[(\omega(k|k-1) + w(k) \cdot \tau)]_i$$
$$A_{i,j}(k) = I \tag{1.32}$$
$$W_{i,j}(k) = \frac{\partial}{\partial w_j}[(\omega(k|k-1) + w(k) \cdot \tau)]_i$$
$$W_{i,j}(k) = [\tau \cdot I] \tag{1.33}$$

The Jacobian H matrix is the partial derivative of the measurement equation $h(x, v)$ with respect to the corrected state vector $(x(k-1|k-1))$. Since the DQ state vector contains only the angular velocity (ω), the H matrix reduces to the partial derivative of the delta quaternion (Δq) with respect to the corrected velocity $(\omega(k-1|k-1))$ (1.34).

$$H(k) = \left[\begin{array}{ccc} \frac{\partial}{\partial \omega_0} \Delta q_i(\omega(k|k-1), \tau) & \frac{\partial}{\partial \omega_1} \Delta q_i(\omega(k|k-1), \tau) & \frac{\partial}{\partial \omega_2} \Delta q_i(\omega(k|k-1), \tau) \end{array} \right] \tag{1.34}$$

The partial derivative of the delta quaternion with respect to angular velocity state was derived in the Q filter derivation (1.22). The DQ measurement equation is linear with respect to measurement noise $(v(k))$, reducing V to the identity matrix (1.35).

$$V_{i,j}(k) = \left[\frac{\partial}{\partial v_j}[h_i(x(k|k-1), v(k))] \right]_{v=0}$$
$$V_{i,j}(k) = I \tag{1.35}$$

1.4.4 QUATERNION PREDICTION

Each of the frameworks support a user-specified prediction time (δ) for maximum flexibility. Future orientation is predicted by assuming that the corrected velocity $\omega(k|k)$ remains constant throughout the prediction interval (Fig. 1.3). The future head orientation $q(k+\delta)$ is estimated as a function of the current quaternion measurement $q(k)$, the angular velocity ω frame time τ and the prediction interval δ (1.36). The function f_P first computes the delta quaternion that occurs if the angular head velocity ω is constant across the prediction interval δ and then applies it to the current quaternion measurement $q(k)$ (1.37).

$$q(k+\delta) = f_P(q(k), \omega(k|k), \delta)) \tag{1.36}$$
$$f_P(q, \omega, \delta)) = \Delta q(\omega(k|k), \delta) \cdot (q(k|k)) \tag{1.37}$$

1.4.5 COMPARISON OF FILTER DESIGN

Each of the frameworks examined uses a multiple stage process to predict orientation (Table 1.1).

Table 1.1: Computational requirements for DQ and Q filters

	Divisions	Additions	Multiplications	Higher Level Functions	Matrix Inverse
Q	12	1612	2092	3	1 (4 × 4)
DQ	18	297	438	3	1 (4 × 4)

Note: Table entries are for one iteration.

The two frameworks examined have widely varied computational requirements due to the complexity of the system and measurement equations (Table 1.2). Approximately the same number of higher-level function calls (sine, cosine, square root) and inverse matrix operations are required by each of the frameworks. The higher-level functions are used in the delta quaternion computation which is common to both frameworks although it appears in different locations in each algorithm.

Table 1.2: Overview of DQ and Q framework methodology

Framework	Pre-Processor	Kalman Filter	Post-Processor
Q	None	Estimate angular head velocity to predict the next quaternion value using a single EKF	Predict future orientation as a function of the current quaternion orientation, head velocity, and the prediction time
DQ	Convert quaternion orientation to delta quaternion	Estimate head velocity using the delta quaternion as the measurement data	A delta quaternion estimating the change in orientation across the prediction interval is applied to the current quaternion measurement to predict future orientation

The single 4 × 4 matrix inverse operation in each framework occurs in the computation of the Kalman gain and fortunately is not affected by the expanded state vector of the Q framework EKF. The Q framework requires substantially more multiplications and additions than the DQ due to the larger state variable. The seven-element state vector of the Q EKF requires three 7 × 7 matrices (A, A^T, and P) in probability covariance calculation. Additionally, the W matrix expands to 7 × 3 and the H expands to 4 × 7. The expanded matrices of the Q framework are each applied multiple times during the Kalman filter prediction-correction process resulting in a five-fold increase of additions and multiplication for the Q as compared to the DQ.

1.5 EXPERIMENTAL ANALYSIS

1.5.1 EXPERIMENTAL DATA

Quaternion head motion data were collected in a simulation of a cockpit VR environment using a Polhemus Liberty AC magnetic tracker operating at a 120 Hz frame rate. The data collection setup consisted of a single Polhemus magnetic sensor mounted on the rear of a headband worn by the test subject. A Polhemus magnetic source was positioned approximately six inches behind the test subject. There was no effort to control the alignment of the sensor in the source frame. Thirteen individual data sets were collected for this experiment, three sets targeting specific head motion categories (benign, moderate and aggressive) and ten additional sets containing the full range of motion expected in a VR cockpit simulation session. Each of the thirteen data sets consists of 10,000 data frames representing 83.33 seconds of continuous data collection. The three motion-specific data sets (tuning data) will be used for filter tuning and performance analysis under specific types of head motion.

The benign motion data set consists of stationary head orientation with smooth, gradual transitions between orientations and is intended to represent targeting and observation activities (Fig. 1.4).

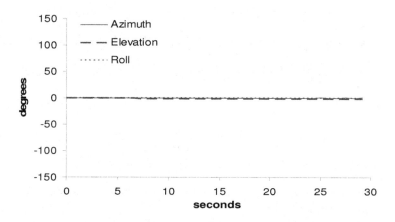

Figure 1.4: Benign head motion data represents semi-stationary activities such as weapons control (first 30 seconds shown as Euler angles in degrees).

The moderate motion data set includes discrete head orientations with smooth transitions at moderate velocities similar to the visual scanning motion a pilot might use (Fig. 1.5). The aggressive data set is included to represent high velocity tracking head movement with rapid starts and stops as would be experienced when a pilot attempts to find a target (Fig. 1.6).

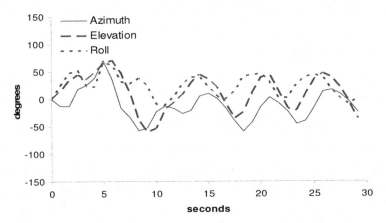

Figure 1.5: Moderate head motion data showing smooth but rapid head motion (first 30 seconds shown as Euler angles in degrees).

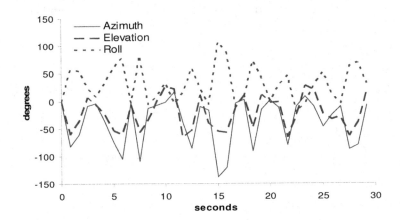

Figure 1.6: Aggressive head motion data (first 30 seconds shown as Euler angles in degrees). Note the very rapid and sometimes erratic motion.

The ten full range motion data sets are intended to be representative of typical head motion during a cockpit simulation session and will be used for performance analysis. The data sets contain intervals of benign, moderate and aggressive motion in pseudo-random order (Fig. 1.7)

1.5.2 TUNING

The Kalman filter uses the process noise covariance and measurement noise covariance to tune the filter for the targeted application. Virtual reality environments are typically custom built in

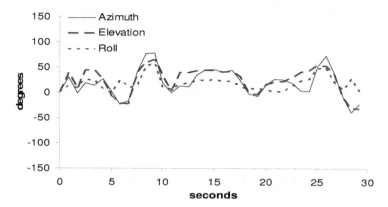

Figure 1.7: Full head motion data is a continuous data capture session that includes a complete range of head motion to closely match simulation session data (first 30 seconds of a typical example shown as Euler angles).

small lots leading to a large variation in how the magnetic source and sensor are positioned in the simulation environment. The two covariance parameters were determined directly from the measured data to allow customization of the filter tuning parameters to each installation. Although this approach does mean that the results are specific to the collected data set, the process is easily repeatable in an installation environment and in fact, could be included in the tracker firmware application. The tuning parameters were derived directly from a composite data set constructed by combining the three tuning data sets (benign, moderate and aggressive) and two of the full motion data sets. This approach was chosen to provide an even weighting of the three categories of head motion while including intermediate data types not represented by the three tuning data sets.

Measurement Noise Covariance

The DQ and Q filters use different measurement data in the correction phase of the Extended Kalman Filter. The DQ filter uses delta quaternion data derived from the quaternion measurement while the Q filter uses the quaternion measurement itself. The measurement noise was defined as the difference between measurement data and a "de-noised" version of the same data. For the Q filter, an estimate of the underlying "noiseless" version of the composite data was created by smoothing with a Gaussian kernel. The smoothed quaternion was then subtracted from the measured quaternion to estimate the measurement noise. The DQ measurement noise is estimated by applying the same technique to delta quaternion data. The delta quaternion measurement data is computed on a frame-by-frame basis from the measured quaternion orientation while the smoothed delta quaternion data is generated from the smoothed data. The measurement noise for the DQ filter is then estimated as the difference between the measured and the smoothed delta quaternion. Using the two-variable quaternion representation (a rotation θ about an axis u) variance of the measurement data for the two filters can be compared (1.38).

$$
z = \begin{bmatrix}
\cos\left(\theta/2\right) \\
u_0 \sin\left(\theta/2\right) \\
u_1 \sin\left(\theta/2\right) \\
u_2 \sin\left(\theta/2\right)
\end{bmatrix}
\tag{1.38}
$$

The z_0 component of the measurement data has a much smaller variance for the DQ than the Q (9.74e-13 vs. 2.55e-07) (Table 1.3).

Table 1.3: Measurement noise covariance for DQ and Q frameworks

	Delta Quaternion Filter	Quaternion (Q) Filter
Variance(z_0)	9.74e-13	2.55e-07
Variance(z_1)	1.21e-08	4.67e-07
Variance(z_2)	3.99e-09	4.10e-07
Variance(z_3)	4.13e-09	3.28e-07

For this experiment, the change in orientation between frames is small due to the high frame rate (120Hz), resulting in a delta quaternion measurement near the identity quaternion. The small changes in rotation θ between frames causes an even smaller variation in the z_0 component of the measurement noise because it is a function of the rotation θ, which has a zero slope for $\theta = 0$. The axis components of the DQ measurement noise (z_1, z_2, and z_3) also have very small variance due to the influence of the sine function with θ near zero ($\sin(0) = 0$). The Q measurement is the total rotation of the current orientation from the origin and is typically not representative of a rotation near zero. Accordingly, the Q measurement noise variance is much larger than the associated DQ values with each of the Q values having similar magnitude due to the averaging effect of the variance calculation.

Process Noise Covariance

The DQ and Q filter both utilize a variation of the angular velocity through the process noise covariance as the driving variable of the Kalman filter prediction step. The Q filter propagates changes in the angular velocity into the quaternion state through application of a delta quaternion (a function of the angular velocity) to the previous quaternion state estimate while the DQ filter uses the delta quaternion itself. In the constant velocity model, the process noise can be modeled as angular acceleration not related to the measurement noise. In this experiment, tuning parameters were derived from measurement data, raising the issue of how to remove measurement noise from the data before estimating the process noise covariance (Table 1.4).

Table 1.4: Process noise covariance for DQ and Q frameworks

	ω_0	ω_1	ω_2
ω_0	9.74e-13	2.55e-07	2.55e-07
ω_1	1.21e-08	4.67e-07	4.67e-07
ω_2	4.13e-09	3.28e-07	3.28e-07

A Gaussian smoothed version of the full range data set was used to provide a "noiseless" quaternion measurement from which to estimate the process noise. A delta quaternion was calculated for each frame of the "noiseless" quaternion and then the angular velocity was estimated by solving the delta quaternion Equation (1.15) using the Levenberg—Marquardt algorithm (LMA). The process noise was estimated from the angular velocity by applying the constant velocity model to the data on a frame-by-frame basis. The difference between the estimated velocity of a given frame and the previous frame estimate was considered to be process noise.

1.5.3   EXECUTION TIME

The single-iteration execution time was measured for each of the frameworks in the MathCAD simulation environment. The iteration time was computed as the average time required to process one frame of data. The DQ framework executed a single pass in 520 microseconds while the Q required 921 microseconds. A tabulation of the number of operations required by each framework (Table 1.1) showed that the DQ provided approximately an 80% improvement in the number of additions and multiplications, but our experimental results showed only a 43.5% improvement.

The less-than-expected improvement in execution time using the DQ is the result of the efficiency of the floating point unit in the simulation host (Pentium 4; 3 GHz). The higher-level functions (sine, cosine, square root) and inverse matrix operations occur at the same frequency in both frameworks, leaving the increased matrix rank of the Q framework as the only difference between the two. Modern floating point units generally can execute one or more multiply/accumulate (MAC) operations in one instruction cycle, reducing the improvement in execution speed.

1.5.4   PREDICTION ACCURACY

Filter performance was rated by comparing the quaternion prediction to the actual time shifted data after conversion to Euler angles (azimuth, elevation and roll). The Euler error for each sample point was computed and then combined to form the RMS average of the compound angle. Error was measured as average error (degrees), overshoot points (as a percentage of totals), overshoot average (degrees) and maximum overshoot (degrees). Overshoot was defined as any error exceeding 1.0 degrees of the composite angle. The two frameworks showed comparable accuracy at the typical prediction time of 50 ms (Table 1.5).

Table 1.5: Framework performance for DQ and Q at 50 ms of prediction

Filter	Benign			Moderate			Aggressive		
	Avg. Err	OS Avg.	OS Max	Avg. Err	OS Avg.	OS Max	Avg. Err	OS Avg.	OS Max
DQ	0.04	0.00	0.00	0.00	0.31	5.22	1.46	2.69	1.11
Q	0.05	0.00	0.00	0.00	0.33	0.58	1.17	1.34	1.77
No Pred.	0.04	0.00	0.00	0.00	1.22	39.73	2.51	5.05	3.79

Notes: All measurements in degrees.

When used with benign motion, there was essentially no error using any of the prediction methods since the orientation did not change appreciably during the prediction interval. Testing under moderate and aggressive motion illustrates the great improvement in prediction that the Q and DQ filters provide as compared to no prediction. For moderate motion the overshoot percentage dropped by 35% and approached a 50% improvement for aggressive motion. The Q filter provided better prediction for moderate motion than the DQ filter. Overshoot was higher for the DQ than the Q during moderate motion (5.22% vs. 0.58%) but average error was not significantly different (0.31 vs. 0.33). Maximum overshoot for the DQ was approximately twice that of the Q for moderate motion (2.69 vs. 1.34) while average overshoots were only slightly higher for the DQ. Overall, the performance data at 50 ms suggests that the DQ filter will handle aggressive behavior better than the Q filter at the cost of performance for moderate motion.

Looking at performance by category as a function of prediction time, we see that both filters had similar performance but with different profiles. Average error of the DQ for moderate motion was slightly better than the Q values when prediction time was reduced below 50 ms but increased significantly above the Q for prediction times greater than 50 ms (Fig. 1.8).

With aggressive data, average error was lower in the DQ than the Q at all prediction times. The overshoot average (Fig. 1.9) and percentage overshoot (Fig. 1.10) was always lower with the Q than the DQ for moderate motion, but the DQ was better with aggressive motion.

Maximum overshoot (Fig. 1.11) showed the Q had better performance during moderate motion but worse performance during aggressive motion.

The DQ filter output is very responsive to changes in angular velocity since these changes directly impact the delta quaternion, which is applied to the measured quaternion for prediction. The Q filter however applies the delta quaternion to the quaternion state which is dependent on not only the quaternion measurement, but also the Kalman gain (per the correction process). Changes in the delta quaternion for the Q filter are not directly reflected in the output, they must propagate through the filter, slowing the response and increasing the error for aggressive motion or increased prediction time. The reduced performance of the DQ with moderate motion is primarily due to larger overshoots since the average error is not significantly different from the

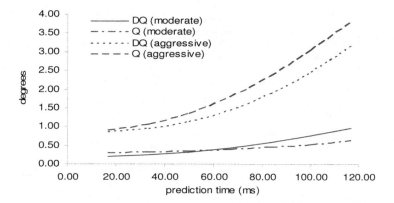

Figure 1.8: Average error for moderate and aggressive head motion as a function or prediction time (compound error in degrees). The DQ performs better with aggressive motion than the Q but is slightly worse when using moderate motion data.

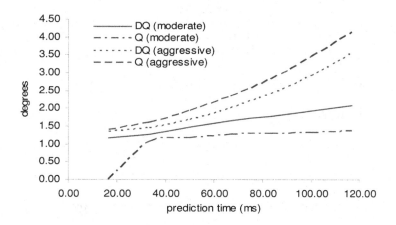

Figure 1.9: Average overshoot vs. prediction time for moderate and aggressive head motion (total overshoot in degrees). The Q performs much better with moderate motion but is much worse with aggressive motion.

Q. The improved responsiveness of the DQ helps it perform better under aggressive motion but also causes it to suffer from increased overshoot during moderate motion.

The DQ and Q filters were also tested with ten different full motion data sets to measure expected performance in a VR simulation environment. All performance measurements of the ten sets were calculated across the combined 100,000 frame sample time (13.88 minutes) to cre-

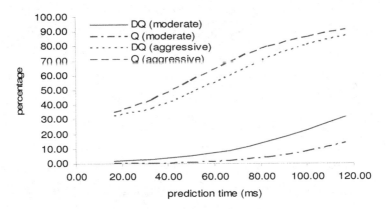

Figure 1.10: Percentage overshoot as a function of prediction time for moderate and aggressive head motion (percentage of sample size).

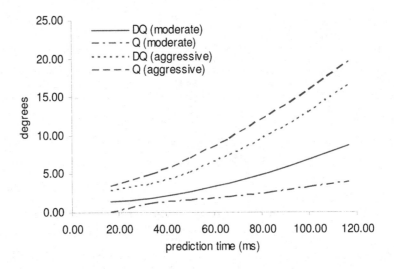

Figure 1.11: Maximum overshoot for moderate and aggressive data as a function of prediction time (shown in degrees). Note that the Q provides the best performance with moderate motion but the DQ is better for the aggressive case.

ate a profile for the DQ and Q filters as a function of prediction time. The DQ filter provided improved performance in all error measurements when using full motion data at any prediction time. The improved aggressive motion performance of the DQ allows it to respond quickly to sudden movements, reducing the average error across the entire simulation interval (Fig. 1.12).

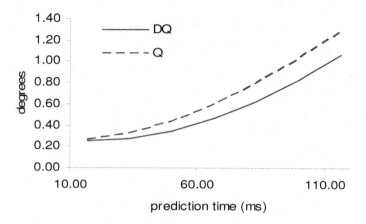

Figure 1.12: Average error vs. prediction time for full motion data shows the DQ outperforming the Q framework for all prediction times.

Overshoot average was slightly improved with the Delta Quaternion Filter (Fig. 1.13).

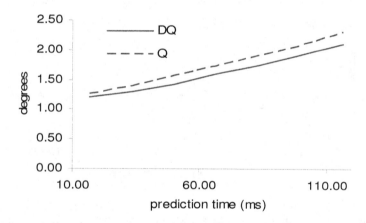

Figure 1.13: Average overshoot was marginally better for the DQ with the full motion data.

Overshoot percentage (Fig. 1.14) and maximum overshoot (Fig. 1.15) were significantly improved with the Delta Quaternion Prediction as compared to the Quaternion.

The results for the full motion simulations suggest that the improved performance of the DQ during aggressive motion is a dominant factor in the overall performance of the prediction process.

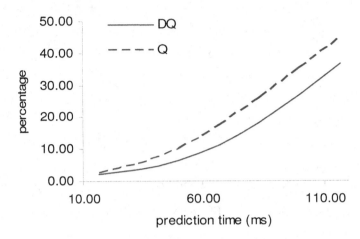

Figure 1.14: The DQ had a lower overshoot percentage than the Q at all prediction times, with almost 10% improvement at 110 milliseconds.

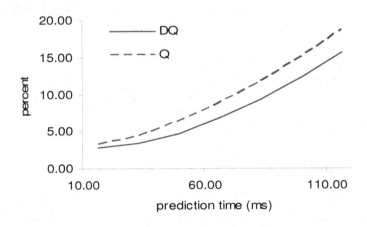

Figure 1.15: Maximum overshoot was significantly improved with the DQ at all prediction times using full motion data.

1.5.5   NOISE PERFORMANCE

The prediction process introduces noise into the quaternion data when it projects the current head velocity forward in time. Small changes in the estimated velocity caused by the prediction-correction behavior of the Kalman filter are amplified by the prediction process. For this experiment, output noise was estimated as the difference between the output data and a smoothed version of itself, expressed in dB. The expectation was that output SNR would drop as the prediction time increased. As shown in Fig. 1.16, the SNR dropped approximately 7 dB when prediction was increased from 0 ms to 120 milliseconds.

The two filters displayed nearly identical noise performance with the DQ filter being slightly better than the Q (Fig. 1.16).

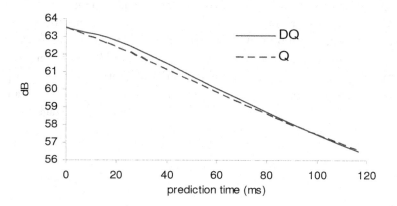

Figure 1.16: Output SNR (dB) as a function of prediction time (milliseconds) was nearly identical for the two frameworks (full motion data).

The DQ displayed increasing output noise as the head motion changed from benign to aggressive (Fig. 1.17).

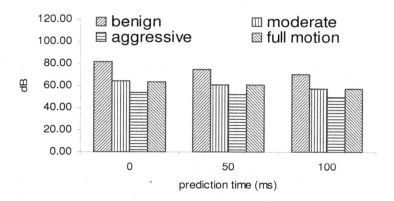

Figure 1.17: DQ framework output SNR (dB) versus prediction time (milliseconds) by motion category (benign, moderate, aggressive and full range head motion). Note that the full motion category has nearly the same SNR performance as the moderate motion one.

The full motion data set provided similar noise performance to the moderate data, suggesting that it is a relatively equal weighting of the three data categories. The 0 ms prediction case indicates that the majority of the change is caused by the tracker, not the prediction algorithm.

1.6   SUMMARY

The Delta Quaternion Filter reduces the computation requirements of quaternion orientation prediction while reducing overshoot. This novel filter provides the performance of the Quaternion filter with a much smaller "footprint" with regard to computation requirements and memory usage. Prediction performance was similar to the Quaternion filter but there was increased error as motion moved toward the aggressive category. The inability of the filter to provide accurate prediction with aggressive motion is a reflection of the wide dynamic range and unpredictable nature of head motion. This is the first stage of DQ development; in stage two we use multiple DQ filters to provide accurate prediction with all motion categories.

CHAPTER 2

Multiple Model Delta Quaternion Filter

2.1 INTRODUCTION

The value of VR/AR in systems using head tracking is directly impacted by the degree of user immersion. Any perceived lag between head motion and scene response causes a loss of immersion that decreases training value [45]. Severe scene lag can disorient the individual, causing dizziness and, in extreme cases, simulation sickness [8–10, 15, 23, 28]. The scene-rendering process in modern VR/AR environments is typically in the range of 50 ms to 100 ms, resulting in significant display lag. An effective method of lag compensation is to predict head orientation using head-tracking data, rendering the next scene ahead of time. Head motion is extremely unpredictable, ranging from stationary pose when studying a scene, to rapid accelerations and decelerations when tracking moving objects. A single motion model cannot accurately handle all types of head motion, resulting in compromised performance [45, 48]. Multiple model estimation can be used to improve head tracking by combining different motion models that target specific types of head motion.

Multiple model algorithms can be divided into three generations: Autonomous Multiple Models (AMM), Cooperating Multiple Models (CMM) and Variable Structure Multiple Models (VSMM) [51]. The AMM algorithm uses a fixed number of motion models operating autonomously. The AMM output estimate is typically computed as a weighted average of the filter estimates. The CMM algorithm improves on AMM by allowing the individual filters to cooperate. The well-known Interacting Multiple Model Estimator (IMME) algorithm is part of the CMM generation. The IMME makes the overall filter recursive by modifying the initial state vector and covariance of each filter through a probability weighted mixing of all the model states and probabilities [53]. The IMME approach was shown to improve performance in high-acceleration conditions but, similar to the modified AMM method, it caused larger overshoot and ringing. The VSMM algorithm builds on the CMM approach by varying the type of models operating in the filter at any given time. Models are dynamically added or deleted from the filter based on their performance, eliminating poorly performing ones and adding candidates for improved estimation.

The Delta Quaternion Filter is implemented in a multiple model framework, the Multiple Model Delta Quaternion (MMDQ), to estimate angular head velocity and acceleration. The rationale behind moving to a multiple model filter is two-fold; first, head motion as too wide a dynamic range for one predictive filter and secondly, the estimation of acceleration in addition to velocity will improve the prediction results. The MMDQ estimates angular head velocity and acceleration from orientation data using an IMME. The IMME was modified to improve overall performance by adding provisions to avoid numerical underflow/overflow conditions and an adaptive transition probability matrix (TPM). The MMDQ uses three extended DQ filters to estimate velocity and acceleration from the change in head orientation expressed as the delta quaternion (Δq). An adaptive prediction algorithm then uses the velocity and acceleration estimates to predict future orientation across a user-specified time interval. This method differs from other EKF-based approaches in that the prediction time is not a multiple of the data rate but can be matched to display lag without consideration of the data rate. The decoupling of the prediction interval from the orientation measurement rate allows the prediction process to make full use of the faster update rate of modern orientation measurement systems.

2.2   RELATED WORK

The author conducted an initial study on using the EKF for head orientation prediction that presented two adaptive approaches [11]. The first adaptive method modified the EKF predicted error covariance to improve tracking performance when head motion changed. Although tracking performance improved, the fading memory algorithm also resulted in increased noise in the predicted orientation. A second adaptive method (R-Adaptive) modified the measurement noise covariance of the EKF in response to the noise level of the predicted orientation. The R-Adaptive successfully controlled the output noise level while improving tracking for benign head motion, but also resulted in increased prediction error when aggressive head motion occurred. The author has previously presented the delta quaternion EKF as a new approach to head orientation prediction [47]. The delta quaternion method removes the quaternion orientation from the EKF, significantly reducing the computation requirements. The study found that the delta quaternion EKF was superior to the well-known quaternion EKF [38, 39] for aggressive head motion but was slightly inferior for moderate head motion. There was no difference between the two approaches for benign head motion.

A modified AMM algorithm was used by Kyger and Maybeck [45] to compensate for latency. Individual filters were designed for look-angle tracking based on First-Order Gauss-Markov Acceleration (FOGMA), Velocity (FOGMV) and Constant Position (CP) models. The three filters ran independently and were reinitialized when divergence was detected. A restart algorithm was added to the AMM framework to keep the individual filter state vectors in the locality of the overall filter state vector, allowing rapid transition between filters as the type of motion changed. The modified AMM filter reduced lag significantly but suffered from increased overshoot and ringing. The filter used one-step prediction to compensate for latency, thus limiting the frame rate to the required prediction time in the application. Additionally, the approach did not provide complete orientation data, choosing to supply look-angle only. Liang et al. developed a head motion prediction method based on Kalman filtering [28]. The proposed system predicted head orientation using a filter model that decoupled the four quaternion components, filtered them independently with separate EKFs, and then recombined them to form the predicted quaternion value. A study of predictive filtering methods was conducted by Rhijn et al. [7]. Their work found that the EKF provided the same performance in typical VR/AR applications as other predictive filtering methods including particle filters and the unscented Kalman filter. Yang et al. studied the use of the EKF in single filter and multiple model frameworks for conflict detection algorithms [46]. Their study found that the single Kalman filter provided some advantages during mode transitions but the multiple model approach performed better overall.

Ali et al. used delta quaternions to control attitude in the Mars Exploration Rover [49]. Their approach estimates the change in orientation and then corrects it using measurement data from a variety of instruments including accelerometers and gyroscopes. Cheon and Kim estimated spacecraft attitude using quaternions [59]. Their work used a magnetometer and gyroscope to estimate quaternion orientation with an unscented Kalman filter. Marins et al. used delta quaternions with Kalman filtering to construct MARG (Magnetic, Angular Rate and Gravity) sensor [35]. A study by Sabatini combined a gyroscope, an accelerometer and a magnetometer to measure orientation for biomedical applications [36]. Each of these studies used angular rate data to estimate quaternion orientation with Kalman filtering. In our study, we estimate angular rate from orientation data, and then use it to predict orientation.

The concept of Delta Quaternion, which hinges on the idea of building an error quaternion using quaternion composition rather than quaternion subtraction, is at the heart of what is known as the Multiplicative Extended Kalman Filter (MEKF) [39, 62, 63]. The MEKF has been used not only to estimate the quaternion, but also to estimate angular velocities as well as typical sensor errors, such as biases, alignments and orientation measurements.

2.3 BACKGROUND

2.3.1 QUATERNIONS AND DELTA QUATERNION

Unit quaternions are popular four-parameter orientation representations with one constraint that avoids the singularities of Euler angles and is more compact than rotation matrices. A quaternion (q) provides a convenient mathematical notation for representing orientation as a rotation (θ) about a unit vector ($\hat{u}$) located in three-dimensional space (2.1) [16, 44, 58].

$$q = \begin{bmatrix} \cos(\theta/2) & \hat{u}\sin(\theta/2) \end{bmatrix}^T \tag{2.1}$$

When constrained to the unit sphere, quaternions provide a unique representation of orientation, but implementation issues cause a sign ambiguity ($\pm q$ is the same rotation). Since this work estimates the change in rotation and applies it in quaternion space, it is not affected by this ambiguity. The rotation component of the quaternion ($cos(\theta/2)$) is forced position to avoid the ambiguity from causing arithmetic problems. The Delta Quaternion (DQ) filter predicts future head orientation from the change in quaternion orientation, computing the change in orientation

(Δq) as a function of the estimated head velocity. To rotate an object, the orientation $q(k)$ of the object is multiplied by the desired change in rotation, i.e., the DQ $\Delta q(k)$ defined as (2.2). Note that the product is determined using a quaternion multiplication ($\otimes$).

$$q(k) = \Delta q(k) \otimes q(k-1) \rightarrow \Delta q(k) = q(k) \otimes (q(k-1))^{-1} \tag{2.2}$$

The DQ filter converts quaternion data to delta quaternions in real time and then applies Kalman filtering, removing the quaternion orientation from the filter state variable and reducing the computational load when compared to quaternion filtering. The average angular velocity between measurements is estimated as an Euler value (azimuth, elevation and roll), and then corrected with the measured change in orientation. The relationship between the delta quaternion and average angular velocity (ω) given by Chou [16] is used to convert Euler velocity to delta quaternions. When acceleration values are used, they are used to modify the average velocity which is then converted to delta quaternions.

2.3.2   EXTENDED KALMAN FILTER

The Extended Kalman Filter (EKF) provides a method of applying the Kalman filter prediction-correction behavior to non-linear systems [13, 31]. In the EKF, the non-linear state equation $f(x(k-1), u(k-1), w(k-1))$ relates the state at time k ($x(k)$) to the previous state ($x(k-1)$) (2.3). Additional parameters in the non-linear state equation are a driving function b (which is not used in this application) and process noise w, where w has the property of the zero-mean white Gaussian noise. The measurement Equation (2.4) relates the state vector ($x(k)$) to the measurement data through the measurement function $h(x(k), v(k))$.

$$x(k) = f(x(k-1), b(k-1), w(k-1)) \tag{2.3}$$
$$z(k) = h(x(k), v(k)) \tag{2.4}$$

In reality, the process noise is not exactly known at time k so the state equation is an approximation ($x(k|k-1)$) of the true next state ($x(k)$) as a function of the previously corrected state ($x(k-1|k-1)$). The notation $x(k|k-1)$ is used to express the state vector at time step k conditioned on the previous state vector at time step $k-1$. Similarly, the measurement function produces an approximation ($z(k|k)$) of the measurement data ($z(k)$) due to the unknown value of the measurement noise v, where v has the property of the zero-mean white Gaussian noise at time k. The governing equation for the EKF state estimate approximates the true state vector ($x(k)$) and the true measurement ($z(k)$) using a Taylor expansion about conditional state ($x(k|k-1)$) (2.5),(2.6).

$$x(k) \approx x(k|k-1) + A(k) \cdot (x(k-1) - x(k-1|k-1)) + W(k) \cdot w(k-1) \tag{2.5}$$
$$z(k) \approx z(k|k) + H(k) \cdot (x(k-1) - x(k-1|k-1)) + V(k) \cdot v(k) \tag{2.6}$$

The *A* and *W* in (2.5) are the Jacobian matrices of partial derivatives of the state Equation (2.2) with respect to the state vector *x* and the process noise *w*, respectively. The true measurement (*z(k)*) relates to the approximate measurement (*z(k|k)*) using the two matrices (*H* and *V*) and the measurement noise *v* (2.5). The *H* and *V* in (2.6) are the Jacobian matrices of partial derivatives of the measurement function *h* with respect to the state vector *x* and measurement noise *v*, respectively.

2.3.3 INTERACTING MULTIPLE MODEL ESTIMATOR

The Interacting Multiple Model Estimator (IMME) has four distinct steps: interaction, filtering, mode probability update, and combination [47, 50, 53, 55, 56]. Figure 2.1 depicts a two-filter IMM estimator where *x* is the system state and *z* is the measurement data. Note that the previous state of each filter is reinitialized by the interaction stage each time the filter iterates.

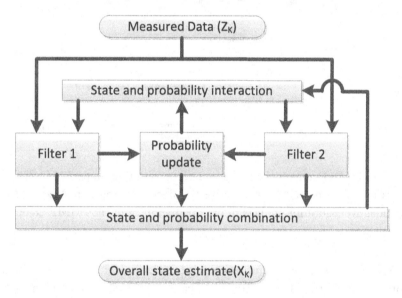

Figure 2.1: The IMME is a four-stage filter that combines different state models into a single estimator to improve performance.

2.4 FILTER DESIGN

2.4.1 MMDQ DESIGN

The MMDQ filter builds upon previous work with the DQ filter [47], improving on the DQ framework by replacing the single EKF with a three-model modified IMME [20] and changing the prediction algorithm to take advantage of the additional resources of the MM state vector. The overall MMDQ filter can be broken into seven separate processes: delta quaternion computation, transition probability matrix update, probability and state mixing, extended Kalman filtering, weighting coefficient computation, state vector combination and orientation prediction (Figure 2.2). The MMDQ filter does not include the quaternion state in the filter state vector, significantly reducing the complexity of the Kalman filters.

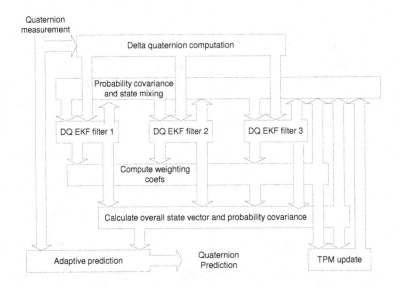

Figure 2.2: The MMDQ expands the DQ approach to use a three-model IMME for head tracking. The IMME is modified to include an adaptive transition probability matrix (TPM) for improved tracking. An adaptive algorithm predicts future orientation from the IMME state estimate and the measured orientation.

The IMME mode-switching process is assumed to be a Markov chain with a known mode transition probability matrix (TPM). The TPM can be estimated from off-line data as a function of the expected sojourn time in each model [51, 54]. Although a fixed TPM can provide good results in most cases, the widely varying nature of a head-tracking application presents large demands on the TPM estimation procedure [41]. For this experiment an adaptive algorithm for TPM estimation operating on an initial estimate will be used. For this discussion equation notation will use subscripts for individual matrix elements ($M_{i,j}$) and bracketed superscript to identity matrix columns ($M^{<i>}$).

A cost-effective method of computing the on-line TPM using a quasi-Bayesian estimator was presented by Li and Jilkov [51]. This method first computes the mixture probability density function (PDF) $g_{i,j}$ for the jth state element of the ith model from the likelihood function (L), the weighting coefficients (μ) and the previous TPM (Π) (2.7). Next, the Dirichlet distribution parameters γ are calculated from the PDF (g) and the previous parameters (2.8).

$$g_{i,j}(k+1) = 1 + \frac{\mu_i(k) \cdot \left(L_j(k) - \left(\Pi^{\langle j \rangle}(k)\right)^T \cdot L(k)\right)}{(\mu(k))^T \cdot \Pi(k) \cdot L(k)} \tag{2.7}$$

$$\gamma_{i,j}(k+1) = \gamma_{i,j}(k) + \frac{\gamma_{i,j}(k) \cdot g_{i,j}(k)}{\sum_j \gamma_{i,j}(k) \cdot g_{i,j}(k)} \tag{2.8}$$

Finally, the new TPM (Π) is computed as the average of the Dirichlet distribution parameters over previous k frames (2.9), which have been modified to prevent any element of the TPM from reaching zero by enforcing a minimum value of 10^{-50}. A zero element in the TPM can produce divide-by-zero exceptions in the implementation of the filter and must be avoided [57].

$$\Pi_{i,j}(k+1) = \max\left[10^{-50}, \frac{1}{k+1} \cdot \gamma_{i,j}(k)\right] \tag{2.9}$$

The initial value of the TPM will be determined through analysis of a dataset that is representative of a typical head-tracking application. Each frame of the data will be identified with a specific motion model. The TPM elements will then be computed as the single step probability of each mode transition. The adaptive computations above are inserted in the IMME structure before the probability mixing stage.

The probability covariance and state vector of each EKF in the IMME are biased toward the overall solution of the IMME before the filters iterate. Each EKF filter is adjusted to the overall solution to prevent filter divergence, keeping the filter state near the operating point of the IMME. The MMDQ modifies this stage by applying a minimum value to the mixing coefficients and weighting coefficients to prevent any value from reaching zero. The recursive nature of the IMME can result in a filter being dropped from use once its weighting coefficient reaches zero [57]. The addition of a lower limit to the mixing process assures that a filter with high error can be effectively removed from the state estimation process without permanently dropping it from the MMDQ.

Head motion is very unpredictable, ranging from benign, stationary pose to erratic, aggressive target tracking. The MMDQ deals with these wide variations by switching between multiple filters; each designed to handle a specific type of head motion. The high measurement rate of the electromagnetic trackers allows for the use of simple motion models such as the constant velocity (CV) and constant acceleration (CA) models. The weighting coefficients are computed using the standard IMME method, computing each coefficient as the product of the previous frame coefficient and the likelihood function. After computation, the weighting coefficients are normalized and a lower bound is applied to avoid zero values that can affect the mixing process. The state vector combination uses the weighting coefficients to generate the overall state vector and probability covariance.

2.4.2   DELTA QUATERNION FILTER DESIGN

Multiple Model approaches are often used to improve prediction by using multiple instances of the same model, each tuned to handle a different type of head motion [48]. We have chosen to use two constant velocity (CV) filters and a constant acceleration (CA) filter, each with different process noise. The high data rate of the simulation environment (120 Hz or more) allows us to use simple motion models such as the CV and CA for head tracking, reducing the complexity of the Kalman filters. The first CV filter will have low-level white noise and will be designed for slow changing and stationary orientation. The second CV filter will have high-level white noise and is intended for moderately changing head orientation. The CA filter will have high-level white noise to handle large changes in acceleration such as starts and stops.

Constant Velocity Filter

The Constant velocity filter uses a state vector (x_{CV}) (2.10) containing the corrected average angular velocity $(\omega(k|k))$ to estimate the delta quaternion Δq.

$$x_{CV}(k) = [\omega(k|k)] \tag{2.10}$$

The CV model state equation $f_{CV}(x, w)$ predicts the next state vector $(x_{CV}(k|k-1))$ as a function of the corrected state vector from the previous frame $(x_{CV}(k-1|k-1))$ and the process noise (w). Since the CV model assumes that velocity does not change between measurements, the estimated velocity $(\omega(k|k-1))$ is a linear function of the corrected angular velocity state $(\omega(k-1|k-1))$, process noise (w) and the frame period (τ) (2.11).

$$f_{CV}(x(k-1), w(k-1)) = \omega(k-1|k-1) + w(k-1) \cdot \tau \tag{2.11}$$

The measurement equation $h(x, v)$ converts the estimated angular velocity to a delta quaternion (2.12). Note that the delta quaternion CV filter has a linear state Equation (2.11) but a non-linear measurement Equation (2.12). The measurement equation is identical in both of our motion models and therefore does not carry a model subscript.

$$h(x(k), v(k)) = \Delta q(\omega(k|k-1), \tau) + v(k) \tag{2.12}$$

The A matrix for the CV model (A_{CV}) is the partial derivative of (1.11) at time k with respect to state (x); this reduces to the identity matrix (2.13). The W matrix for the CV model (W_{CV}) is the partial derivative of (2.11) with respect to process noise (w), reducing to the frame time (τ) multiplied by the identity matrix (2.14). The V matrix is the partial derivative of (2.12) with respect to measurement noise (v), evaluated at the current state. Both of our EKF implementations assume that the measurement noise is additive which reduces V to the identity matrix (2.15).

$$A_{CV}(k) = \left[\frac{\partial}{\partial x_{CV}} f_{CV}(x_{CV}(k-1), w(k)) \right]_{w=0}$$

$$A_{CV} = I \tag{2.13}$$

$$W_{CV}(k) = \left[\frac{\partial}{\partial w} f_{CV}(x_{CV}(k-1), w(k)) \right]_{w=0}$$

$$W_{CV}(k) = I \tag{2.14}$$

$$V(k) = \left[\frac{\partial}{\partial v} h(x(k|k-1), v(k)) \right]_{v=0}$$

$$V(k) = I \tag{2.15}$$

The H matrix at time step k is the partial derivative of (2.12) at time step k with respect to the state variable (x). Expressing H as three-column vectors, the general form is a function of the estimated delta quaternion ($\Delta q(\omega(k|k-1), \tau)$), the estimated angular velocity ($\omega(k|k-1)$) and the sample period (τ), all at time k (2.16).

$$H_{CV}(k) = \frac{\partial}{\partial \omega}[h(x_{CV}(k|k-1), 0)]$$

$$(H_{CV}(k))^{\langle i \rangle} = \begin{bmatrix} -\frac{\tau}{4} \cdot \Delta q_{i+1} \\ \frac{\omega_0}{\omega^T \omega}(\tau \omega_i \Delta q_0 - \Delta q_{i+1}) + \delta_{i,0} \frac{\Delta q_{i+1}}{\omega_i} \\ \frac{\omega_1}{\omega^T \omega}(\tau \omega_i \Delta q_0 - \Delta q_{i+1}) + \delta_{i,1} \frac{\Delta q_{i+1}}{\omega_i} \\ \frac{\omega_2}{\omega^T \omega}(\tau \omega_i \Delta q_0 - \Delta q_{i+1}) + \delta_{i,2} \frac{\Delta q_{i+1}}{\omega_i} \end{bmatrix} \tag{2.16}$$

where $\omega = \omega(k|k-1)$ $\Delta q = \Delta q(\omega(k|k-1), \tau)$

and $\delta_{i,j}$ is the Dirac function $\delta_{i,j} = \begin{cases} 1 & if \quad i = j \\ 0 & if \quad i \neq j \end{cases}$

Constant Acceleration Filter

The constant acceleration (CA) filter models the changes in quaternion orientation as incremental accelerations between measurements [52]. The state vector of the CA filter at time k ($x_{CA}(k)$) contains the corrected angular velocity ($\omega(k|k)$) and corrected angular acceleration ($\alpha(k|k)$) (2.17).

$$x_{CA}(k) = \begin{bmatrix} \omega(k|k) & \alpha(k|k) \end{bmatrix}^T \tag{2.17}$$

The CA state equation $f_{CA}(x, w)$ expresses the predicted velocity ($\omega(k|k-1)$) as the sum of the velocity state ($\omega(k-1|k-1)$) and the product of the angular acceleration state ($\alpha(k-1|k-1)$), the frame time (τ) and the process noise (w) (2.18). The predicted angular acceleration ($\alpha(k|k-1)$) is the sum of the current acceleration state (α(k-1|k-1)) and the product of the process noise (w) and the frame time (τ). The CA filter uses that same measurement equation as the CV filters (2.12).

$$f_{CA}\left(x\left(k-1\right),w\left(k-1\right)\right) = \begin{bmatrix} \omega + \alpha\tau + w\frac{\tau^2}{2} \\ \alpha + w\tau \end{bmatrix} \tag{2.18}$$

The A and W Jacobian matrices for the CA filter (A_{CA}, W_{CA}) can be derived by inspection from the expanded form. The A_{CA} matrix (2.19) and W_{CA} matrix (2.20) are derived separately for the CA filter but the V matrix is unchanged since we are using the same measurement model (2.12).

$$A_{CA}\left(k\right) = \frac{\partial}{\partial x} f_{CA}\left(x_{CA}\left(k-1\right),0\right)$$

$$A_{CA} = \begin{bmatrix} I & \tau \cdot I \\ 0 & I \end{bmatrix} \tag{2.19}$$

$$W_{CA}\left(k\right) = \frac{\partial}{\partial w} f_{CA}\left(x_{CA}\left(k-1\right),0\right)$$

$$W_{CA} = \begin{bmatrix} \frac{\tau^2}{2} \cdot I & \tau \cdot I \end{bmatrix}^T \tag{2.20}$$

The H matrix for the CA filter (H_{CA}) contains the partial derivatives of the measurement Equation (2.12) with respect to each of the state variables (2.10). The general form of H_{CA} uses the same three-column vectors of (2.16) but with the CA model used to compute the predicted angular velocity. Three additional columns containing the partial derivatives with respect to acceleration are then appended (2.21).

$$H_{CA}\left(k\right) = \begin{bmatrix} \frac{\partial}{\partial \omega} h\left(x_{CA}\left(k|k-1\right),0\right) & \frac{\partial}{\partial \alpha} h\left(x_{CA}\left(k|k-1\right),0\right) \end{bmatrix} \tag{2.21}$$

The three partial derivatives of (2.12) with respect to acceleration (α) are functions of estimated angular velocity (ω(k|k-1)), estimated delta quaternion ($\Delta q(\omega$(k|k-1),t)) and the sample period (τ). The H_{CA} matrix can be expressed in a compact form by noting that the partial derivative with respect to angular acceleration (α) differ only in a term of τ from the partial's with respect to angular velocity (w) (2.22).

$$H_{CA}\left(k\right) = \begin{bmatrix} I & \tau \cdot I \end{bmatrix} \cdot \begin{bmatrix} \frac{\partial}{\partial \omega} h\left(x_{CA}\left(k|k-1\right),0\right) \end{bmatrix} \tag{2.22}$$

2.4.3   ORIENTATION PREDICTION

Future orientation is estimated by computing the delta quaternion (Δq) expected during the prediction interval (δ) from the corrected state estimate (x(k|k)) and applying it to the current quaternion state estimate $(q(k))$. The prediction interval is divided in $N=\delta/\tau$ time steps and the velocity for each of the N steps is computed by applying the CV and CA models to the current MMDQ state estimate $(x(k|k))$. The weighted average of each of the N steps is computed using the model weighting $(u_{CV}(k), u_{CA}(k))$ from the MMDQ (2.23). The delta quaternion of each step is computed and applied recursively to determine the quaternion orientation (q^E) after each complete time step in the prediction interval (2.24). The predicted quaternion orientation for time step k $(q^P(k))$ is computed from the final version of q^E, including any additional partial step time in the prediction interval (2.25).

$$\omega(k+n|k) = u_{CV}(k) \cdot \omega_{CV}(k) + u_{CA}(k) \cdot (\omega_{CA}(k) + \alpha_{CV}(k) \cdot \tau) \quad (2.23)$$

$$q^E((k+n|k),(n \leq N)) = \Delta q(\omega(k+n|k),\tau) \otimes q(k+n-1|k) \quad (2.24)$$

$$q^P(k) = \Delta q(\omega(k+n|k),(N \cdot \tau - \delta)) \otimes q^E(k+n|n=N) \quad (2.25)$$

$$N = trunc\left(\delta/\tau\right) \quad n = 1 \dots N$$

2.5 EXPERIMENTAL RESULTS

Head motion data was collected using a Polhemus Liberty AC magnetic tracker to provide measured data for the experiment. The experiment setup used a single sensor attached to the rear of a helmet with the magnetic source rigidly mounted approximately 0.2 m from the sensor. Each of the collected datasets contains 100,000 sequential head orientation samples collected at a 120 Hz measurement rate.

A quaternion orientation dataset was collected for three specific head motion categories (benign, moderate and aggressive motion). The three motion categories were chosen to correlate with those used by Kyger [45] in his experiment with multiple model head orientation prediction. In their experiment, Kyger and Maybeck assembled these three motion categories from data captured during simulator missions with experienced pilots at Armstrong laboratories. For this experiment head orientation data was created for each of these categories to closely match that of the Kyger experiment by carefully controlling head motion while recording head orientation.

In Figure 2.3 the normalized histogram of each of the datasets is seen to occupy a separate region of the angular velocity range. The benign motion dataset has a distribution that is sharply defined with very little acceleration content as would be expected when the pilot studies a stationary object. The moderate data set has a wider range of values that represents smooth motion as the pilot scans the airspace. The aggressive dataset distribution is very broad and represents fast, erratic motion and has a similar maximum value (14.6 radians/sec2; Table 2.1) to that used by Kyger.

Two additional datasets (Motion 0; Motion 1) were taken with a full range of head motion for performance evaluation (Motion 0; Motion 1). Each of the full motion datasets features a complete range of head motion data from benign, stationary pose, to wildly aggressive tracking

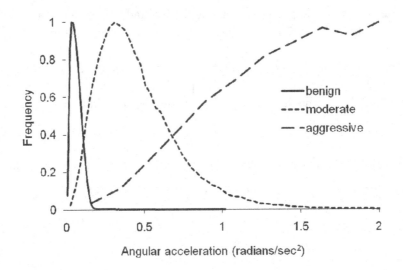

Figure 2.3: A histogram of the benign, moderate and aggressive motion sets normalized to the same frequency scale. Note that three motion models define specific ranges of angular acceleration with overlapping regions. The aggressive motion histogram is only partially shown.

Table 2.1: Angular head motion by dataset

Dataset	Velocity (radians/s)			Acceleration (radians/s^2)		
	mean	stdev	Max.	mean	stdev	Max.
Benign	3.8e-3	2.0e-3	0.08	0.06	0.03	1.02
Moderate	0.32	0.03	0.76	0.47	0.25	3.27
Aggressive	1.05	0.22	2.83	4.00	3.07	18.6
Motion 0	0.27	2.92	1.02	0.98	3.90	10.8
Motion 1	0.36	2.82	2.92	1.20	4.37	14.6

motions, at random intervals similar to that expected in a simulation environment. A histogram of the angular acceleration in each of the two motion datasets (Figure 2.4) shows that these datasets are predominantly moderate and benign motion with short random bursts of aggressive motion.

The angular acceleration in the two full motion datasets (Motion 0 and Motion 1) have a large standard deviation (~ 4 radians/s^2) and a small mean value (~1 radian/s^2) indicating the head experiences short bursts of high acceleration (Table 2.1).

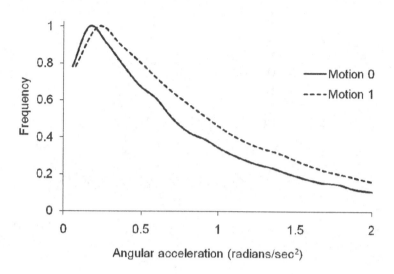

Figure 2.4: A histogram of the angular head acceleration for the two full motion datasets shows the head is generally experiencing moderate or benign motion. Note that the "tail" of each histograms is not shown to emphasis the peak near 0.25 radians/sec².

2.5.1 MMDQ CONFIGURATION

The full motion data sets (Motion 0; Motion 1) were evaluated to determine if the maneuvering index [64] of the collected datasets requires the IMM. The maneuvering index (λ) is the ratio of the standard deviation of the process noise to the standard deviation of the measurement noise. The CV process noise of each point was computed using a 12-point sliding window centered on the point. The DQ filter uses a delta quaternion measurement to estimate angular velocity and therefore a direct comparison the process and measurement noise is not possible. To calculate the maneuvering index, the delta quaternion data was converted to an angular velocity measurement from which the measurement noise was derived. The full-motion datasets contain points with a maneuvering index greater than 0.5 (Fig. 2.5), indicating that the IMM will provide improved tracking over a single EKF [64].

The distribution of λ shows that a single EKF has adequate bandwidth for more than 75% of the samples (Table 2.2) but there will be outliers that will be difficult for the filter to track. Given the narrow band of the maneuvering index, a three-filter MMDQ may not be necessary. To investigate this, a two-model (MMDQ2) was implemented in addition to the originally proposed three-model version (MMDQ3). A CV model EKF (DQEKF-CV) and CA model EKF (DQEKF-CA) were also implemented to provide a performance comparison to the single stage DQ.

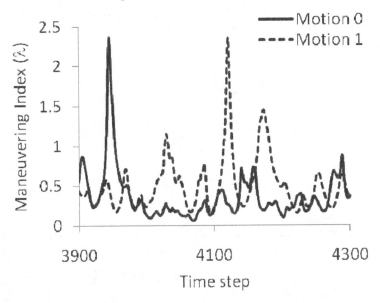

Figure 2.5: A segment of the Motion 0 and Motion 1 datasets that display a large maneuvering index value. The large dynamic range of the data as it changes from benign motion to aggressive motion cannot be handled by a single EKF without large estimation errors.

Table 2.2: Maneuvering index (λ) distribution for full motion datasets

	1st Quartile	Median	3rd Quartile	98 percentile
Motion 0	0.16	0.26	0.48	1.07
Motion 1	0.17	0.28	0.49	1.07

The three-filter MMDQ (MMDQ3) will use a CV filter for $(0.25 \leq \lambda < 0.75)$ (CV1), a CV filter for $(0.5 \leq \lambda < 1.0)$ (CA1) and a second CA filter for $(\lambda > 1.0)$ (CA2). The small number of points served by the CA2 filter in the MMDQ3 configuration raises the question of whether a two-filter configuration provides similar performance. The two-filter configuration (MMDQ2) that uses a CV filter tuned for midrange $(0.25 < \lambda < 0.75)$ (CV1) and a CA filter for the moderate motion $(0.25 \leq \lambda < 0.75)$ will also be implemented. The individual filters are tuned at the midpoint at the assigned maneuvering index range (Table 2.3).

The initial value of the transition probability matrix (TPM) was estimated by assuming that all state changes were the result of a single step Markov chain. The assigned ranges of λ were then used to assign a filter to each frame of the Motion 0 dataset.

Table 2.3: MMDQ filters vs. maneuvering index (λ)

Filter Type	MMDQ3	MMDQ2
CV1	0.5	0.5
CV2	1.0	0.5
CA2	8.0	0.5

2.5.2   TPM INITIALIZATION

The probability of a transition from filter i to filter j ($\Pi_{i,j}$) is the number of transitions from i to j ($N_{i,j}$) in the dataset divided by the total transitions from filter i ($\Sigma_j N_{i,j}$) (2.26). A TPM was generated for the IMM3 (2.27) and MMDQ2 (2.28) configurations using the described process.

$$\Pi_{i,j} = N_{i,j} / \sum_j N_{i,j} \tag{2.26}$$

$$\Pi_{IMM3} = \begin{bmatrix} 0.970 & 0.029 & 0.001 \\ 0.069 & 0.917 & 0.014 \\ 0.001 & 0.250 & 0.749 \end{bmatrix} \tag{2.27}$$

$$\Pi_{IMM2} = \begin{bmatrix} 0.971 & 0.029 \\ 0.065 & 0.935 \end{bmatrix} \tag{2.28}$$

The probability of transitions between the low process noise CV1 filter and the CA1 filter in the MMDQ3 (2.27) were initially set to zero based on the procedure outlined above. It was reasoned that the high data rate of the tracker (120 Hz) was eliminating direct transitions between these two filters and forcing them to transit through the CV2 filter. Experimentation with the TPM showed that allowing transitions from CV1 to CA1 improved tracking by reducing the mode transition time at the onset of accelerations. A small transition probability in the CV1/CA1 location was sufficient to allow CV1/CA1 transitions.

2.5.3   MEASUREMENT NOISE

The measurement noise ($v(k)$) is common to all three filters since the measurement equations are identical. To find $v(k)$ for a dataset, a smoothed version of the data set is subtracting from the measurement. This method was chosen so as to include dynamic errors of the tracker in the computation. Stationary measurement data originally taken for this process was found to have very little noise (<-90 dB) and was not representative of the measurement noise with motion data. For this experiment the individual components of the delta quaternion are assumed to be independent variables, allowing the standard deviation of the measurement noise (2.29) to be used instead of the complete covariance matrix. These are typical values used for the experiments, the actual values were determined during the simulations to support the use of multifold cross validation.

$$\sigma_V = \begin{bmatrix} 2.09e-06 & 0.22e-03 & 8.22e-05 & 7.92e-05 \end{bmatrix}^T \tag{2.29}$$

2.5.4 PROCESS NOISE

The process noise for each filter was determined from an assigned maneuvering index (λ) that represents the range of motion the filter is expected to cover. For the MMDQ filters, the values from Table 2.3 were used, centered in the assigned range of λ. For the EKF filters, the maneuvering index was set at $\lambda = 0.5$ for both the CV (DQEKF-CV) and the CA (DQEKF-CA) versions of the DQ filter. The EKF filter tuning is biased toward moderate motion to provide better performance during aggressive motion.

The standard deviation of the process noise (σ_w) was found by applying the measurement noise (σ_v) and time step to λ. Note that the equation for the CV filter model process noise (2.28) is slightly different than that for the CA filter model (2.29).

$$\sigma_{wCV} = \lambda \cdot \sigma_v / \tau \tag{2.30}$$

$$\sigma_{wCA} = \lambda \cdot \sigma_v / \tau^2 \tag{2.31}$$

The final tuning values for each of the filters are row vectors with each vector containing σ_v values for the azimuth, elevation and roll components of the angular velocity (in that order) (Table 2.4). These tuning values were determined by reviewing simulation results to find a near optimal result.

Table 2.4: MMDQ process noise filter tuning values

	CV1	CA1	CA2
MMDQ3	[1.48, 0.63, 0.72]	[353, 152, 175]	[708, 304, 350]
MMDQ2	[1.48, 0.63, 0.72]	[353, 152, 175]-	–
EKF-CV	[1.48, 0.63, 0.72]		
EKF-CA	–	–	[176, 76, 86]

The final MMDQ2 values have the CV1 filter identical to the individual DQEKF-CV and a high process noise CA filter. The MMDQ3 uses the same tuning for the CV1 and CA1 filter, adding the CA2 filter for very aggressive motion. Notice that the DQEKF-CA uses a much lower process noise than the MMDQ2-CA1. The high gain of the CA1 filter in the MMDQ precludes its use as a standalone filter like the DQEKF-CA due to compromised performance with low acceleration data like the benign motioned dataset.

2.5.5   ANGULAR VELOCITY ESTIMATION

The performance for each of the four filters was evaluated using a 10-fold cross validation process with a 10K sample validation interval and a non-overlapping 90K sample training interval. For the full motion data sets (Motion 0 and Motion 1) the filters were tuned from the training set and performance was measured using the results obtained by running the validation data through the filter. The motion specific datasets (benign, moderate and aggressive) used a similar approach except that tuning was determined by a training interval in the Motion 0 dataset. This approach provided for the evaluation of filter performance for the specific motion classification while tuning the filter for full motion data.

The DQEKF-CV had the lowest maximum error with all three of the motion-specific datasets but had higher median error for benign and aggressive motion, suggesting that the CV motion model is "smoothing" the velocity curve, cutting the min/max values while loosely following the normal waveform. This behavior is expected since the lower dynamic range of the single CV filter cannot respond quickly to changes in velocity, resulting in a response similar to a sliding window average. The EQEKF-CA filter had the best overall results with full motion data with a low median and maximum error. This CA filter is more responsive than the CV, allowing it to closely follow the true velocity. However, the higher gain of the filter is problematic during benign motion where it has the largest errors or all four filters.

The MMDQ filters have varying behavior based on filter type (2 vs. 3 filters) and motion type. The MMDQ2 has good performance with the benign motion set, but the worst performance with the moderate and aggressive sets. The MMDQ3 had the good results for all four datasets, including the full motion dataset. This pattern indicates that three filters in the MMDQ3 are allowing it to rapidly change motion models to remain converged on the velocity waveform.

Table 2.5: MMDQ velocity estimation error by motion category

	Full Motion			Benign Motion			Moderate Motion			Aggressive Motion		
	mean	median	max	mean	median	max	mean	median	max	mean	median	max
MMDQ2	24.3	18.5	207	3.60	3.00	40.6	16.7	14.9	79.3	40.8	32.3	295
MMDQ3	23.0	17.1	205	3.63	3.01	40.6	15.4	13.8	74.3	38.5	31.5	295
DQEKF-CV	22.3	15.6	233	4.05	3.49	38.8	13.4	11.8	66.8	42.6	35.1	259
DQEKF-CA	23.9	17.4	218	5.34	4.51	47.4	16.5	14.2	89.5	43.6	35.8	299

Note: All values are in milliradians/s.

The MMDQ3 filter provided the best overall tracking performance of any of the four filters. With the combination of a CV filter for benign to moderate motion and a CA filter for moderate to aggressive motion, the MMDQ provides the same or better results than any of the other filters. The addition of a third filter to the MMDQ structure for the MMDQ3 provided little improvement in the aggressive data set results as compared to that of the MMDQ2. The lack of improvement with the CA2 filter is mostly likely due to the limited dynamic range of our aggressive motion dataset. In their experiment with multiple model prediction, Kyger [45] used an aggressive motion dataset with angular acceleration approaching 35 radians/s^2 while our aggressive dataset had a maximum acceleration of 18 radians/s^2. An examination of the model weighting in the two MMDQ filters (Figure 2.6) shows that the MMDQ3 is using the CA2 model for high accelerations that use the CA1 model in the MMDQ2. The MMDQ3 is mixing the CA1 and CA2 models for these intervals while the MMDQ3 is relying on the CA1 filter alone. Based on these results, the MMDQ2 would be the better choice for our application since it has all the performance of the MMDQ3 without the additional computational overhead. However, a data set with higher acceleration (like that used by Kyger) may require more process noise than the CA1 can provide, requiring a switch to the MMDQ3.

2.5.6   PREDICTION PERFORMANCE

The MMDQ filters and DQEKF-CA filters had very similar prediction errors across all data sets, with the DQEKF-CA having slightly better results in terms of median error. Display lag compensation is most concerned about overshoot performance, as this kind of error is what causes the "swimming" effect that occurs with large display lag. Looking at overshoot maximums (OS Max.) we see that the MMDQ filters provide the best overall performance. The MMDQ2 tuning was provided the best results for aggressive motion (Fig. 2.7) and the best overshoot performance for all datasets (Tables 2.6, 2.7, 2.8 and 2.9). The MMDQ3 had the best results with moderate motion (Table 2.8).

The difference in performance between the MMDQ2 and MMDQ3 is due to tuning and motion model selection. For the MMDQ2 we chose a CV/CA combination tuned with the same value as the equivalent DQ filter. For the MMDQ3, we chose a CV/CV/CA combination with the two CV filters tuned for benign and moderate motion. The MMDQ3 CA filter was tuned for highly aggressive motion and did not seem to be a factor in the experiment. It should be pointed out that the aggressive motion dataset does not contain the large accelerations of the Kyger experiment. The datasets used by Kyger had accelerations as high as 35 radians/sec^2 while the aggressive dataset had a maximum of 18.5 radians/sec^2. The lack of this extreme level of acceleration may present different results with the filter.

2.5.7   COMPUTATIONAL REQUIREMENTS

The MMDQ contains two (MMDQ2) or three DQEKF filters so we would expect it to consume additional bandwidth when compared to the DQEKF. As shown in Table 2.10, the MMDQ imposes a significantly larger computation load than the DQEKF in both the two- and three-filter versions. The MMDQ2 required more than three times the execution time than the DQEKF while the MMDQ3 required more than four times that of the DQEKF. However, the execution time of the MMDQ is small enough that it can easily be included in the firmware of a typical orientation tracker. For example, the MMDQ3 was added to the Polhemus Liberty tracker used

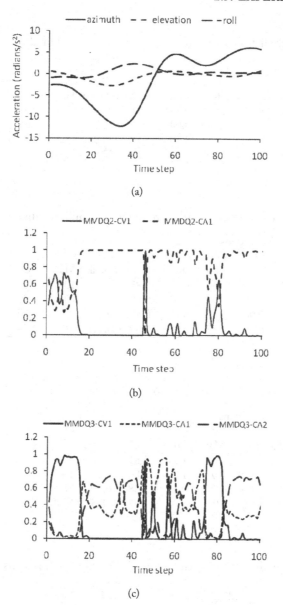

(a)

(b)

(c)

Figure 2.6: A segment of the Motion 0 dataset acceleration illustrates mode switching in the MMDQ filters. Here we see: (a) a plot of angular head acceleration, (b), the mixing weights for MMDQ2 and (c) model mixing weights for MMDQ3. Note that the model switch with smooth curves in the MMDQ3 chart due to the overlapping tuning of the three filters. The MMDQ2 is generally in one model or the other due to the wide spacing of the filter tuning ranges. The MMDQ3 switches in smooth curves, mixing two or more filter outputs due to the overlapping tuning ranges of the filters.

Table 2.6: MMDQ prediction error with full-range motion (milliradians)

	50 ms				75 ms				100 ms			
	median	OS median	OS Max	stdev	median	OS median	OS Max	stdev	median	OS median	OS Max	stdev
MMDQ2	2.34	18.9	23.7	2.73	4.91	21.9	59.0	6.46	8.34	28.0	107	11.7
MMDQ3	2.60	19.1	33.4	3.04	5.28	22.6	67.6	6.65	8.75	25.2	107	11.4
DQEKF-CV	2.75	20.8	40.3	3.79	5.49	23.3	75.2	7.70	9.00	25.9	115	12.8
DQEKF-CA	1.99	20.6	33.0	2.66	4.36	22.9	67.0	6.10	7.58	25.1	106	10.7

Table 2.7: MMDQ prediction error with benign motion (milliradians)

	50 ms				75 ms				100 ms			
	median	OS median	OS Max	stdev	median	OS median	OS Max	stdev	median	OS median	OS Max	stdev
MMDQ2	0.45	0.00	5.57	0.42	0.85	0.00	10.8	0.80	1.30	17.5	17.5	1.24
MMDQ3	0.43	0.00	5.57	0.42	0.85	0.00	10.8	0.80	1.30	17.5	17.5	1.24
DQEKF-CV	0.45	0.00	5.57	0.42	0.85	0.00	10.8	0.80	1.30	17.5	17.5	1.24
DQEKF-CA	0.40	0.00	4.33	0.35	0.80	0.00	9.51	0.74	1.28	0.00	15.8	1.2

Table 2.8: MMDQ prediction error with moderate motion (milliradians)

	50 ms				75 ms				100 ms			
	median	OS median	OS Max	stdev	median	OS median	OS Max	stdev	median	OS median	OS Max	stdev
MMDQ2	1.94	0.00	10.0	1.17	3.81	17.5	17.5	2.29	6.19	19.3	28.2	9.9
MMDQ3	1.98	0.00	8.72	1.19	3.86	0.00	16.2	2.33	6.25	19.2	26.7	3.78
DQEKF-CV	1.99	0.00	9.38	0.12	3.88	0.00	17.4	2.41	6.27	20.0	28.3	3.88
DQEKF-CA	1.60	0.00	10.9	1.08	3.32	18.5	19.4	2.11	5.53	20.0	30.6	3.46

Table 2.9: MMDQ prediction error with aggressive motion (milliradians)

	50 ms				75 ms				100 ms			
	median	OS median	OS Max	stdev	median	OS median	OS Max	stdev	median	OS median	OS Max	stdev
MMDQ2	3.80	19.5	21.8	2.73	9.00	22.4	61.0	6.97	16.3	25.6	117	13.1
MMDQ3	5.20	19.8	34.0	4.09	11.1	23.6	80.5	9.15	19.1	27.6	143	16.0
DQEKF-CV	6.45	21.3	48.3	5.52	12.9	24.6	100	11.2	21.5	29.2	168	18.7
DQEKF-CA	4.57	19.6	33.6	3.81	10.1	23.7	78.3	8.75	17.9	26.9	139	15.5

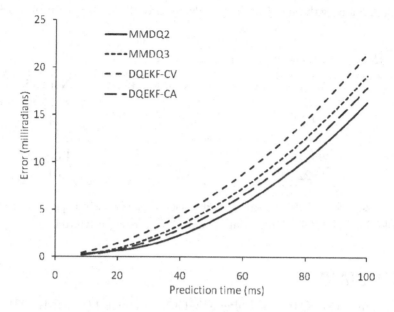

Figure 2.7: The mean error for aggressive motion plotted as a function of prediction time indicates that the MMDQ2 filter has best performance with aggressive motion.

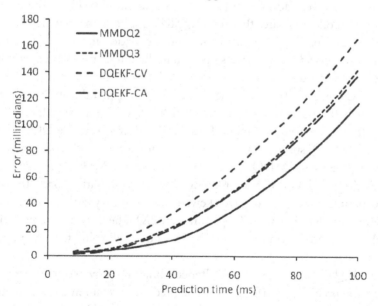

Figure 2.8: The maximum error vs. prediction shows that the MMDQ2 provides the lowest over-shoots or all the filters. The MMDQ3 has performance similar to the DQEKF-CA.

in this study, maintaining the standard 240 Hz data rate of the tracker while improving prediction performance.

Table 2.10: MMDQ computational requirements

Filter	Cycle Count	Execution Time (u sec)	Normalized Bandwidth requirement
MMDQ2	29572	295.7	3.11
MMDQ3	44360	443.6	4.66
EKF-CV	9512	95.1	1.00
EKF-CA	9622	96.2	1.01

Notes: Cycle counts are for a single iteration of the filter and were measured by implementing each algorithm in "C" on an Analog Devices ADSP-21161N floating point DSP operating at 100MHz.

2.6   SUMMARY

Two versions of the MMDQ (the two-filter MMDQ2 and three-filter MMDQ3) were compared to the original DQEKF-CV and a new version using the CA motion model (DQEKF-CA). The DQEKF-CV had the most error of four filters due to the limited dynamic range of this design. The DQEKF-CA however had excellent results for a single stage filter. The maneuvering index for the CA motion model indicates that the DQEKF-CA has a much wider dynamic range than the DQEKF-CV. Comparing the maneuvering index equation for the CV (2.28) and CA (2.29) motion models shows an additional time step term in the numerator. When working with a small time step of 8.33 ms, the additional term greatly reduces the maneuvering index for a system as compared to the CV model. The experimental results confirmed the ability of the DQEKF-CA to be consistently better performing than the DQEKF-CV. Comparing the DQEKF-CA to the two MMDQ filters, there were several performance measures were the DQEKF-CA had better results than the MMDQ, specifically the mean error for benign and moderate motion.

Comparing the two MMDQ filters, we see that the MMDQ2 provides the same level of performance as the MMDQ3 with a 30% reduction in computation load. The limited dynamic range of the experimental data may have skewed this result. Kyger [21] used head motion data with a much larger acceleration range (approximately 2X). The lack of extremely high acceleration data limited the expected dynamic range of the input data, allowing for better optimization with the limited range datasets.

The MMDQ2 filter with a CV/CA filter combination provided that best trade-off of computational load and prediction performance. The prediction requirement for the tracker has been fully addressed with the development of the DQEKF and MMDQ series of filters.

CHAPTER 3

Interpolation Volume Calibration

3.1 INTRODUCTION

The use of magnetic trackers in simulation environments provides an ideal platform for head tracking but has problems in the presence of conductive or ferrous materials. The magnetic tracker uses a dipole field model to measure position and orientation (PnO), calculating the mutual inductance between a magnetic source and a pickup coil sensor. The magnetic field generated by the tracker creates eddy currents on the surface of nearby conductive materials that interfere with tracker operation. Ferrous materials couple into the magnetic field and distort it in the region nearby. The effect of conductive and ferromagnetic materials on magnetic trackers was conducted by Nixon et al. [67]. Position error was confirmed to vary as a fourth order function of the transmitter separation distance. Distortion effects were seen to be highly dependent on distance, a relatively small increase in the distance between the receiver and metal objects reduced error substantially. The effect of conductive and ferrous materials can be mitigated through careful control of the simulation environment but often cannot be eliminated. In these cases, calibration methods are used to correct the tracker measurement based on a mapping of the tracking volume. The mapping operation measures the magnetic field in the tracking volume, capturing the relationship between distorted and true data for the compensation algorithm.

Mapping a tracking volume is generally accomplished through use of a mechanical fixture that precisely locates sensors with known PnO in the tracking volume, allowing the measured PnO to be associated with a mechanical measurement. The mechanical devices must be constructed of non-conductive, non-ferrous materials to avoid additional distortion and can range from simple building block type assemblies that place sensors at known PnO, to motorized equipment that reduce data collection time. These devices are expensive to manufacture and require trained technicians to operate, resulting in high cost to the user.

The Interpolation Volume Calibration (IVC) mapping system was designed to provide field mapping of a distorted environment and generate a LUT of the distortion in the mapped volume. The system combines inexpensive equipment and a new processing algorithm to facilitate the data collection process by inexperienced personnel. This new method requires two pieces of mechanical equipment (in addition to the tracker) to collect data, an interpolation fixture and a mapping fixture, both constructed from non-conductive, non-ferrous material. The interpolation fixture is used to create a volume in the map where we can solve for the true PnO from the collected data and the fixed geometry of the fixture. The mapping fixture is used in conjunction with the interpolation volume to collect a data cloud containing measured field data with known PnO. The data cloud is converted to a uniform grid look-up table (LUT) that contains the magnetic field correction required to obtain PnO in a distorted environment. The LUT can be directly applied to field compensation or used to construct a PnO compensation LUT.

3.2   PREVIOUS WORK

The performance of AC magnetic trackers in distorting environments has been investigated for a wide range of applications. The distortion can be corrected using any of several approaches including management of the tracking volume, single sensor-based compensation or multiple sensor compensation. Management of the tracking volume involves carefully selecting the location of the magnetic sensor and source to minimize distortion. This approach is appropriate for low distortion environments that have some flexibility on the placement of the sensor and source assemblies. Application with higher levels of distortion or physical constraints that require specific source/sensor configurations must use a distortion compensation approach. The compensation schemes can be categorized into two groups: single sensor and multiple sensor compensation.

3.2.1   MANAGING THE TRACKING VOLUME

Careful placement of the magnetic source and sensor in the target environment can be used to reduce the effect of conductive and ferrous materials. A study conducted by Wagner et al. [84] found that magnetic trackers were not ill-suited to computer-aided surgery if a careful selection of instruments based on size and material was conducted. Milne, Chess and Johnson et al. [82] conducted a survey of magnetic tracker accuracy and its susceptibility to distorting materials. The authors found that the worst case distortion occurred when the distorter was placed next to the sensor as opposed to being near the source. Hummel et al. [77] evaluated the performance of a miniature sensor, concluding that careful control of the materials and geometries in the tracking volume can significantly reduce tracking errors. Birkfellnor et al. [75] conducted a series of experiments with electromagnetic tracking systems in surgical environments and found that careful control of distorting materials was required to achieve adequate performance. A comparison of magnetic trackers was conducted by Hummel et al. [77] to determine their suitability for image-guided surgery. Test results indicated that magnetic trackers were suitable for a surgical environment if proper precautions are taken to minimize sources of distortion.

3.2.2   SINGLE SENSOR COMPENSATION

Traditional approaches to distortion compensation use an off-line calibration process to create a mathematical function that is used to correct the PnO data from each sensor. The tracker is used to collect data in the distorted environment and determine the relationship between corrected and measured PnO. Applications such as surgery that require a high degree of accuracy and precision often use magnetic tracking in distorted environments with single sensor calibration. Kindratenko [85] conducted a survey of calibration methods for electromagnetic trackers. In each of the reviewed methods, measurement data was collected at known locations (usually a uniform grid) to capture the relationship between distorted and true data. A polynomial function or an LUT was then constructed to provide error correction for measurement data. In one cited work, a neural network was used to generate the correction factor but no experimental results were provided. A description of several distortion compensation schemes is presented by Raab et al. [74] including polynomial and LUT approaches. Ikitis, Brederson, Hansen and Hollerbach [72] categorize calibration techniques into three categories: analytic (polynomial), global interpolation and local interpolation.

An eddy current compensation system using multiple frequencies to detect characteristics of the distorting material was proposed by Jones et al. [87]. In their work, the group estimated the effect of eddy currents based on the ratio of the in-phase and quadrature components of measured magnetic field. Their method is effective for frequencies in the low audio range but suffer from range limitations due to reduced inductive coupling between the source and sensor.

Another method of eddy current compensation was proposed by Jones and Khalfin [90] that uses additional sensors in a known geometry. These "witness sensors" are used to determine the effective PnO of the magnetic source based on their known relative PnO to the physical PnO of the source.

3.2.3 POLYNOMIAL FUNCTION METHODS

Polynomial function calibration constructs a multivariate polynomial that generates either true PnO or PnO corrections from distorted PnO input. The tracking volume is mapped and the data processed off-line to create polynomial coefficients that are used at run time. A polynomial correction algorithm was used by Nakada et al. [83] in a two-step compensation scheme for laparoscopic surgery. A rigid hybrid optical/magnetic tracker was first used to map the tracking volume just before surgery. Data was collected by moving the sensor tip around the tracking volume in a freehand motion for approximately 30 seconds. The collected data was then used to select one of four polynomial correction functions (1st through 4th order). Polynomial calibration works very well when distortion is smoothly spread through the tracking volume but is not as effective when an abrupt, localized non-linearity occurs in the mapped space.

3.2.4 LOOK-UP-TABLE METHODS

LUT calibration techniques provide a highly localized correction factor that is very successful in applications requiring high precision. The LUT is generally constructed on a regular grid with small spacing (38.1 mm for example) and can accurately provide correction of random field points through interpolation of the grid point values. In a second paper on electromagnetic tracker calibration, Birkfellner et al. [76] studied the use of a hybrid optical/magnetic tracker in a surgical environment. The surgical environment was first mapped with the hybrid tracker in a uniform grid to create a correction LUT. During operation, the optical position measurement from the hybrid tracker was used to access correction factors for the magnetic tracker stored in an LUT. Experimental results showed a significant improvement in accuracy of the magnetic tracker to an average error of 2.8 mm as compared to 4.6 mm for the uncompensated case.

A calibration technique using a look-up table was developed by Day, Murdoch and Dumas [69]. The system collected data on a regular grid in the target environment and computed a position and orientation correction for each grid point. A look-up table (LUT) was then constructed, containing the distortion present at each of the grid points. The system used linear regression to estimate the true position by comparing the measured data to the sum of the true position and the distortion for that position from the LUT.

Another LUT-based correction method was proposed by Ghazisaedy et al. [71]. Data was collected using an ultrasonic device on a regular grid and an LUT of corrections was constructed. The method corrected large errors in reported position but did not significantly improve on small errors. A comprehensive analysis of LUT calibration including data collection and several interpolation algorithms is presented by Jayaram et al. [79]. The authors found that the LUT correction method dramatically improved tracker accuracy but that the improvement varied based on the interpolation method used. Implementation and accuracy issues for LUT correction of magnetic tracking systems was presented by Livingston and State [80]. In their study, the authors determined that position accuracy could be improved by nearly 80% in their experiment although they had difficulty with orientation compensation.

An orientation correction scheme that uses a uniform XYZ grid in the tracking space to record orientation errors was proposed by Ochoa-Manorga et al. [86]. The quaternion measurement data was collected on a uniform grid and then used with a quasi-linear interpolation to generate quaternion corrections. Their method significantly improved tracking accuracy in mildly distorted environments but had difficulty in highly distorted fields.

3.2.5 MULTIPLE SENSOR TECHNIQUES

Multiple sensors in the same tracking environment can be used to add constraints to the system and improve calibration techniques. Feuerstein and Vogel et al. [81] presented a new method of distortion compensation using a hybrid optical/magnetic tracker in laparoscopic surgical applications. They use multiple magnetic sensors mounted at various positions on a laparoscopic probe and a single optical sensor on the exposed end to provide compensated magnetic position measurements without an LUT.

A detailed investigation of errors produced by electromagnetic tracking systems was conducted by Frantz [65]. The authors developed procedures to determine the accuracy and repeatability of electromagnetic tracking systems. Of specific interest in their work was the determination that multiple sensors with rigid positioning can be used to determine the accuracy of the tracking system.

Hagemeister [66] proposed a quick method of determining the coefficients required for a polynomial correction of position and orientation data. A rigid body with multiple sensors in a known PnO relative to each other was used to collect data in the tracking volume. The authors use the multiple sensor measurements in pairs to determine the change in error between each pair of sensors and relate it to the gradient of the polynomial correction function. The technique significantly reduced large errors in the tracking volume without the effort required for a formal mapping.

A method of calibrating a volume for single coil magnetic sensors was proposed by Wu and Taylor [68]. The algorithm used multiple sensors attached to a rigid object in varying orientations to characterize the operating volume. An off-line LSE technique was then used to construct a polynomial correction for the sensor in the targeted environment.

Wang and Jiang [70] propose a novel position compensation scheme that combines multiple sensors in a fixed geometry. Their sensor is constructed of six magneto-resistive (MR) sensors with a single sensor mounted on each surface of a 1cm square cube. Sensors mounted on opposing sides of the cube are paired to produce two position measurements with a known relative distance. When used in the tracking environment, the position of the center of the cube is calculated as the weighted sum of the sensor pairs.

3.3 BACKGROUND

3.3.1 QUATERNIONS

A unit quaternion (q) provides a convenient mathematical notation for representing orientations and rotations (θ) about a unit vector ($\hat{a}$) (3.1).

$$q = \left[\begin{array}{cc} \cos(\theta/2) & \hat{a} \cdot \sin(\theta/2) \end{array} \right]^T \tag{3.1}$$

When constrained to the unit sphere, quaternions provide an effective method for manipulating orientation information, but care must be exercised to deal with the sign ambiguity ($\pm q$ is the same rotation). workaround for the sign ambiguity is to require one of the quaternion components (q_0 for example) to always have a positive magnitude. Note that the quaternion multiplication is a special matrix operation ($\otimes$) involving non-linear functions. The quaternion is computed in its compact four-element form and expanded to a 4×4 matrix for multiplication operations [4, 16, 47]. Quaternion orientation is the preferred representation of orientation in the tracker mathematics due to its compact form and lack of singularities.

3.3.2 AC MAGNETIC TRACKING

An AC electromagnetic tracker determines position and orientation by generating a three-dimensional AC magnetic field and measuring the mutual inductance between sensor/source pairs. The magnetic field is created with a source assembly constructed of three concentric, orthogonal coils using time or frequency multiplexing to differentiate between source coil windings. The magnetic sensors are constructed similarly to the source assemblies using three concentric orthogonal coils. The tracker measures the voltage induced on each of the sensor coils and normalizes the data to the source/sensor assemblies, operating frequency and various physical constants. The data is represented by a 3×3 matrix signal matrix (S) containing the normalized field measurements for each of the 9 source/sensor coil pairs. Each row of S represents the signal received by a specific winding of the sensor while each column is the received signal for a particular winding of the source. An algorithm presented by Jones [73] provides a simplified relationship between the signal matrix and PnO (2.31). The measured signal matrix (S) is the product of the transposed sensor rotation matrix in the source frame (T), and the un-rotated dipole field ($f_D(\cdot)$). Note that several physical constants, the magnetic field frequency and several other parameters have been removed from the equation through the application of the source/sensor calibration.

The dipole field Equation (3.3) calculates the signal matrix based on the outer product ($< rr^T >$) of the position vector (r). Equation (3.3) will be referred to as the *un-rotated* signal matrix since it assumes that the sensor is un-rotated in the source frame, with the position vector describing the offset between the sensor magnetic moment and the source magnetic moment.

$$S = (T)^T f_D(r) \tag{3.2}$$

$$f_D(r) = \frac{1}{\|r\|^3} \left(\frac{3 \cdot \langle rr^T \rangle}{\|r\|^2} 3 - I \right) \tag{3.3}$$

The continuous magnetic field used in AC magnetic trackers causes eddy current induction in surrounding conductive materials. The induced eddy currents create secondary magnetic fields that distort the primary (dipole) field created by the tracker. As shown in Fig. 3.1, the current in the source windings ($I_0 e^{i\omega t}$) induce emf in the sensor windings (ϵ_2^P) through mutual inductance M_{02} and an emf in the conductor loop (ϵ_1^P) through mutual inductance M_{01} [28]. The emf induced in the conductor loop results in an eddy current ($I_1 e^{i\omega t}$) which creates a secondary magnetic field that induces an emf (ϵ_2^S) in the sensor windings through mutual inductance M_{12}. The total emf induced on the sensor windings causes current flow ($I_2 e^{i\omega t}$), generating a voltage proportional to the magnetic field strength at the sensor position.

Without the secondary magnetic field, the sensor winding voltage is a function of the position of the sensor in the source frame (3.1). With the secondary field present, the dipole equation no longer applies since the voltage at the sensor is a function of both the primary and secondary magnetic fields. The measured signal matrix ($S(k)$) is now the sum of the primary field ($f_D(r(k))$) and secondary fields ($G(r(k))$) rotated by the transpose of the sensor rotation ($(T)^T$) in the source frame (3.4).

$$S = (T)^T (f_D(r) + G) \tag{3.4}$$

The effect of the error in the signal matrix due varies with the spatial arrangement of the source, sensor and conductor. Since the eddy current is essentially a single axis source winding,

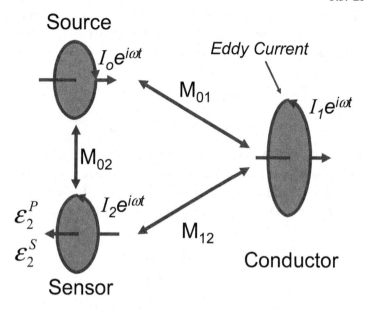

Figure 3.1: A simple circuit representation of how eddy currents affect an AC electromagnetic tracker. The induced current in the conductor produces a secondary magnetic field that induces a secondary emf in the sensor which causes tracking errors.

the mutual inductance is dependent on the relative range and orientation. The mutual inductance between the source and conductor (M01) is primarily a function of range since the source has three orthogonal coils, each operating with similar magnetic moments. The mutual inductance between each individual source winding and the conductor loop is determined by the orientation of the conductor in the source frame. The resulting secondary field will contain components from each source winding. The mutual inductance from the conductor to the sensor (M12) is also primarily a function of range between the sensor and the conductor but the coupling to the individual sensor windings is heavily dependent on the relative orientation of the sensor. The conductor eddy current will induce different magnitudes of the secondary field into each sensor winding based on the relative orientation of the sensor to the conductor loop. The overall effect of the secondary field is a non-linear PnO error factor that smoothly varies with both position and orientation as the sensor moves.

The impact of the secondary field on the tracker PnO varies with each type of measurement performed. The sensor range is inversely proportional to the magnitude of the magnetic field. The magnitude of the measured field at time step k is found from the measured signal matrix by taking the dot product of the measured signal matrix (B(k)) (3.5). The diagonal in B is the voltage induced in each sensor winding for both the primary and secondary fields. Note that there is no orientation factor in the equation; the range is only affected by the magnitude of the secondary field emf induced on the sensor winding. The sensor range (R(k)) can now be found from the trace of B(k)—independent of the sensor rotation (3.6).

$$B\left(S\left(k\right)\right) = S(k)^T S\left(k\right) \tag{3.5}$$

$$R\left(k\right) = \sqrt[6]{6/tr(B(k))} \tag{3.6}$$

The position of sensor at time step k $(r(k))$ is found using a three-step process from the previous position solution $(r(k-1))$ and $B(k)$. The position unit vector $(\hat{u}(k))$ is found as the normalized product of $(B(k))$ and the previous position $(r(k-1))$. The new position $(r(k))$ is then computed as the product of the range at time k $(R(k))$ and the unit vector at time k $(\hat{u}(k))$ (3.5).

$$\hat{u}\left(k\right) = B\left(k\right) \cdot r\left(k-1\right)$$

$$r\left(k\right) = R\left(k\right) \cdot \hat{u}\left(k\right) \tag{3.7}$$

The components in the measured signal matrix $(S(k))$ due to the secondary field have a greater effect on the position calculation due to the orientation-dependent coupling of the individual sensor windings and the conductor loop. With the position now determined, the orientation $(q(k))$ can be solved as the product of the unrotated dipole solution (3.3) and the inverse of the measured signal matrix (3.8). The orientation is the most effected measurement since it is dependent on the accuracy of the position solution to compute the un-rotated dipole field; even small errors in the position will cause large swings in the calculated orientation.

$$T\left(k\right) = f_D\left(r\left(k\right)\right)\left(S\left(k\right)\right)^{-1} \tag{3.8}$$

The secondary field generated by eddy currents distorts the range, position and orientation measurements when it sums into the dipole field at the sensor position. Compensation methods found in publication correct the position and orientation in distorted tracking volumes through polynomial function-based correction factors. The correction factors are developed by mapping the volume to determine a relationship between the measured PnO and the true PnO. As shown in (3.7) and (3.8), the error in a particular PnO due to the secondary field is orientation dependent. Polynomial functions are accurate at the specific PnO of the sensor when the measurement was made but have increasing error as the sensor PnO is varied. In practice, these compensation techniques are usable in mild distortion but become increasingly inaccurate as the distortion component of the measured signal matrix increases.

The more accurate compensation algorithms require the construction of an LUT containing either field data or PnO measurements. When using a field data LUT, the tracker uses the tabulated data to determine the correct PnO using any one of a number of different algorithms that match the LUT to the measured field. The accuracy of field data compensation is only limited by the ability to locate the correct data in the LUT. There are several different methods of using a PnO LUT, each requiring different kinds of data. The table can contain position corrections, the correct PnO or some kind of coefficient for a PnO compensation algorithm. Generally, PnO compensation is only usable while the measured PnO is relatively close to the correct value. As the distortion level increases, the measured PnO quickly becomes very non-linear and indexing into a LUT table with it is problematic. Field-based compensation schemes work well at all distortion levels since they skip the PnO computation as a means of indexing LUT data. However, they require precision equipment to accurately determine the sensor PnO when the data is taken since tracker PnO measurements are inaccurate.

3.4   FIELD MAPPING USING IVC

The IVC system creates a secondary field LUT from field data that can be used for LUT-based field compensation or to develop data for other compensation methods. The system collects field data using two different fixtures to form a "data cloud" of field points that are then interpolated to construct the LUT. The data points are interpolated to produce a uniform grid LUT of the secondary magnetic field.

The system is built on the premise that LUT compensation removes the requirement for absolute measurement accuracy in the mapping process. In our approach, we allow the user to specify a measurement reference frame through placement of a small fixture in the tracking environment. All sensor PnO measurements are made relative to the fixture pose, not the magnetic source reference frame. The declared fixture PnO is not required to accurately reflect the true fixture PnO in the source reference frame, eliminating the need for precise alignment of the source and measurement reference frames.

The user determines the LUT alignment and the measurement reference frame when the interpolation fixture is placed in the mapped volume. The interpolation fixture also defines a small volume in which the PnO of a sensor at an arbitrary field point (in the volume) can be determined. Once the volume (the interpolation volume) is defined, the mapping fixture is used to collect field data throughout the mapped volume. The PnO of the mapping fixture sensors are extrapolated from the interpolation volume through the known, fixed geometry of the fixture sensors.

3.4.1   INTERPOLATION VOLUME

The interpolation volume creates an area of the mapped volume where we can determine the PnO of a sensor without using mechanical measurements. The interpolation fixture takes the form of a cube with a sensor on each vertex (Fig. 3.2). The fixture is then placed in the environment at known PnO to determine the secondary (distortion) field at each point. The measured field for any point inside the interpolation volume can be estimated as the sum of the interpolated secondary field and the dipole field. Combining the field estimation with a cost function and minimization process, the PnO of arbitrary points inside the interpolation volume can be determined.

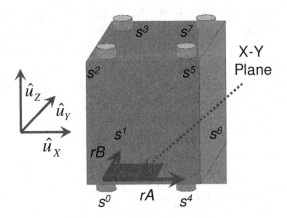

Figure 3.2: The interpolation fixture shown with sensors placed to construct a cube with a sensor at each vertex. The averaged positions of sensors 0, 2 and 4 are used to define the X-Y plane of the fixture reference frame.

To create the interpolation volume, a small fixture that positions a sensor on each vertex of a cube is constructed from non-conductive, non-ferrous material. Each sensor is assigned a number from one to seven based on its XYZ position on the cube. This assignment scheme simplifies the placement of the fixture in the tracking environment in a manner conducive to the interpolation scheme. The sensor orientation can be arbitrary since it is measured when the fixture is calibrated. The fixture calibration measures the PnO of the fixture sensors with the tracker in a non-distorting environment. The fixture can be placed at an arbitrary position in the source reference frame while a large set (> 1000) sensor PnO measurements are captured. The collected data is averaged to remove noise and then used to determine the relative position of the sensors on the fixture.

The PnO of each sensor n on the fixture is determined from the measured sensor positions by constructing a fixture reference frame. In this discussion we will use r_n to denote the position vector r of sensor n in the source (global) reference frame. The X-Y plane of the fixture reference frame is defined on the fixture by two vectors r^A (3.9) and r^B (3.10) along orthogonal edges of the cube. The r^A vector is along the X-axis of the proposed reference frame while the r^B vector is along the Z-axis. The X-axis unit vector $\hat{u}^X$ is defined as parallel to vector r^A (3.11).

$$r^A = r_4 - r_0 \tag{3.9}$$

$$r^B = r_2 - r_0 \tag{3.10}$$

$$\hat{u}^X = r^B - r^A / \left\| r^B - r^A \right\| \tag{3.11}$$

The Z-axis unit vector $\hat{u}^Z$ is defined as the cross product of vectors r^A and r^B (3.12). The Y-axis unit vector$\hat{u}^Y$ is the cross product of the X and Z unit vectors (3.13).

$$\hat{u}^Z = r^A \times r^B \Big/ \left\| r^A \times r^B \right\| \tag{3.12}$$

$$\hat{u}^Y = r^A \times r^B \Big/ \left\| r^A \times r^B \right\| \tag{3.13}$$

The fixture reference frame T^I (3.14) can now be used to transform sensor positions between the fixture and source reference frames. The sensor positions in the fixture reference frame are transformed from the source (global) reference frame measurements using T^I (3.15). The relative quaternion orientation of sensors on the fixture is computed by converting the rotation matrix (T^I) to quaternions (q^I) (3.16), where we use the notation "$TtoQ(T)$" to indicate a standard conversion function. The quaternion frame orientation (q^I) is then used to rotate the measured quaternion of the sensor 0 into the fixture reference frame (3.17).

$$T^I = \begin{bmatrix} \hat{u}^X & | & \hat{u}^Y & | & \hat{u}^Z \end{bmatrix} \tag{3.14}$$

$$r_n^P = T^I \left(r_n - r_0 \right) \tag{3.15}$$

$$q^I = TtoQ \left(T^I \right) \tag{3.16}$$

$$q_n^P = \left(q^I \right)^{-1} \otimes q_n \tag{3.17}$$

After calibration the interpolation fixture is placed in the tracking environment with a known PnO. The initial orientation is set by aligning the fixture reference frame with the source reference frame in a repeatable manner and declaring the fixture orientation as aligned. Inaccuracies in the initial PnO appear as additional distortion of the field and will be compensated for in the estimation process as constant bias in the secondary field. Theoretically, the system should be able to handle any bias but these errors needlessly increase the non-linearity of the system and can impact the accuracy of secondary field estimates.

When the fixture has been placed in tracking volume, the user-provided fixture PnO determines the PnO of each fixture sensor through the fixture reference frame. The position of each interpolation fixture sensor n in the source frame (r_n^F) is the sum of the sensor 0 position (r_0) and the relative sensor position (r_n^P) rotated by the fixture rotation matrix (T^I) (3.18). The fixture sensor quaternion orientation is the quaternion product of the fixture orientation (q^I) and the relative sensor orientation (q_n^P) (3.19).

$$r_n^F = \left(r_0 + T^I r_n^P \right) \tag{3.18}$$

$$q_n^F = q^I \otimes q_n^P \tag{3.19}$$

Once the interpolation fixture is aligned to the source frame, J frames of field data ($J >$ 100) are collected and averaged to estimate the expected value of the measured signal matrix for each fixture sensor (S_n).

Each sensor is at an arbitrary orientation in the source reference frame (Fig. 3.3). The secondary field estimates must be aligned to the source reference frame for the interpolation process (i.e., they all must be at the same orientation with respect to the source).

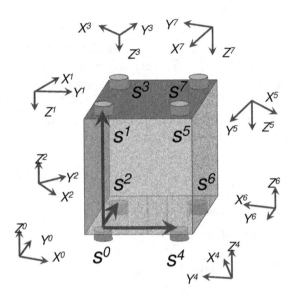

Figure 3.3: The measured field at each fixture sensor is at a different orientation in the source reference frame and must be rotated into alignment with the source before it is used for interpolation. Using the calibration data, a rotation matrix for each sensor is developed to rotate the signal matrix into the fixture reference frame.

The secondary field for each fixture sensor (G_n) is determined from the un-rotated averaged signal matrix by subtracting the primary field at the specified fixture position. Rearranging (3.2), the secondary field for sensor n is computed as the difference between the un-rotated average signal matrix and the un-rotated dipole field (3.20). Note that the orientation of sensor n in the source reference frame (q_n^F) is used to remove the sensor rotation from the measured signal matrix (S_n) before the dipole field is subtracted.

$$G_n = QtoT\left(q_n^F\right) \cdot S^n - f_D\left(r_n^F\right) \qquad (3.20)$$

Using the known secondary field values at the cube vertices (G) and the known fixture sensor positions (r^F), the secondary field can be estimated at any point in the cube volume using a tri-linear interpolation function ($f_I(\cdot)$) (3.21).

$$g(r) = f_I\left(r, r^F, G\right) \tag{3.21}$$

The interpolation function estimates the signal matrix of the secondary field without rotation in the source reference frame as a function of sensor position. Since (3.3) computes the unrotated primary field as a function of position, the measured signal matrix can be calculated for arbitrary field points inside interpolation volume as a function of position. Combining the dipole equation with the secondary field interpolation, the measured field is estimated as a function of position (r) and quaternion orientation (q) (3.22). The quaternion orientation (q) is converted to a rotation matrix (T) using a standard conversion function denoted as "$QtoT(\cdot)$."

$$f_S(r, q) = (QtoT(q))^T\left(f_D(r) + f_I\left(r, r^F, G\right)\right) \tag{3.22}$$

A cost function minimization process based on the dot product of the signal matrix $(B(S))$ is used to solve the sensor position in the interpolation volume (3.5). The signal matrix S is the product of the sensor rotation and the un-rotated signal matrix at the sensor position, taking the dot product $(B(S))$ results in a rotation invariant measurement of the signal matrix at the sensor position. Using position as the independent variable, the cost function minimizes the difference in $B(S)$ between the estimated signal matrix (3.22) and measured signal matrix (S) (3.23). The estimated signal matrix is computed using (3.22) with an arbitrary orientation (here the identity quaternion for q^I is used).

$$\arg\min\left[B(S) - B\left(f_S\left(r, q^I\right)\right)\right] \tag{3.23}$$

The sensor orientation is solved using a second cost function that minimizes the difference between the measured signal matrix (S) and a rotation of the estimated signal matrix at the position r (3.24). Using (3.22), the position is kept constant at the value found with (3.23) while the quaternion orientation is the independent variable. The cost function normalizes the estimated orientation each time it iterates to keep the quaternion solution on the unit sphere.

$$\arg\min\left[(f_S(r, q) - S\right] \tag{3.24}$$

The interpolation volume provides the ability to determine the PnO of a single sensor in a distorted environment. A second fixture is now designed to collect field data at multiple positions in the tracking volume.

3.4.2   MAPPING FIXTURE

The mapping fixture is used in conjunction with the interpolation volume to measure the magnetic field. The fixture is similar to a wand or stick constructed from a rigid non-conductive, non-ferrous material that has multiple sensors mounted at regular intervals from one end to the other. The mapping process keeps one sensor constantly inside of the interpolation volume while the other end of the fixture is slowly moved through the mapped volume.

The mapping fixture is constructed with sensors mounted at regular intervals down its length at arbitrary orientation (Fig. 3.4).

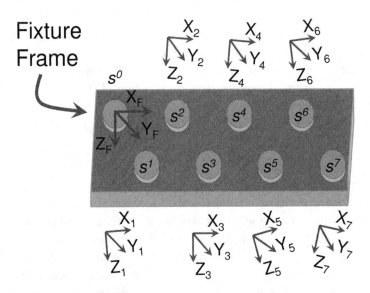

Figure 3.4: The mapping fixture has multiple sensors (eight in this case) mounted on a non-conductive material. The fixture sensors are located at regular intervals down the length of the material with less than 38.1 mm of offset to provide with arbitrary orientation. A fixture reference frame is constructed using the sensor 0 orientation. The sensor PnO in the fixture frame is determined by calibrating the fixture in a non-distorting environment.

The fixture sensor PnO is measured through a calibration procedure similar to that used for the interpolation fixture. The appliance is placed in a non-distorting environment and a large data set (> 100) of PnO measurements for each sensor is captured. The collected PnO is averaged and then used to determine the relative PnO of each sensor on the mapping fixture.

The position of fixture sensor n relative to sensor 0 (r_n^M) is computed as the difference between the average position of sensor n (r_n') and sensor 0 (r_0') during the mapping fixture calibration (3.25).

$$r_n^M = r_n' - r_0' \tag{3.25}$$

The orientation of sensor n in the mapping fixture frame (q_n^M) is computed as the quaternion product of the sensor n average orientation (q_n') and the inverse of the sensor 0 average orientation (q_0') (3.26).

$$q_n^M = q_n' \otimes \left(q'_n\right)^{-1} \tag{3.26}$$

The mapping fixture uses the orientation of sensor 0 to calculate the orientation of all the other sensors based on the fixture frame; the orientation of sensor 0 is determined through the interpolation process. To determine each sensor PnO in the source reference frame during the data collection process, the estimated sensor 0 PnO is applied to the relative PnO developed in (3.25) and (3.26). The mapping fixture reference frame at time step k ($T^M(k)$) is defined as the sensor 0 orientation (3.27). Sensor n position in the source reference frame at time step k is the sum of sensor 0 position ($r_0(k)$) and the relative position of sensor n (r_n^M) rotated by the fixture rotation ($T^M(k)$) (3.28). The orientation of sensor n at time step k ($q_n(k)$) is the relative sensor orientation (q_n^B) multiplied by the orientation of sensor 0 ($q_0(k)$) (3.29).

$$T^M(k) = QtoT(q_0(k)) \tag{3.27}$$

$$r_n(k) = r_0(k) + T^M(k) \cdot r_n^B \tag{3.28}$$

$$q_n(k) = q_n^M \otimes q_0(k) \tag{3.29}$$

The mapping fixture uses the orientation of sensor 0 to calculate the orientation of all the other sensors based on the fixture frame while the orientation of sensor 0 is determined through the interpolation process. Errors in the estimation process are propagated through (3.28) and (3.29) and have an increasing effect on the estimated PnO of the mapping fixture sensor as the offset from sensor 0 increases. To improve the accuracy of the interpolation estimate, a method of estimating the errors of each mapping sensor PnO estimate must be developed. The cost function solver used to find orientation inside the interpolation volume rotates the estimated signal matrix to find the best match with the measured signal matrix. The un-rotated signal matrix (a signal matrix for a sensor aligned to the source reference frame) is essentially three magnitude values, one for each sensor winding. When a sensor is rotated, the three vectors are mixed as a function of the sensor orientation in the source reference frame. The rotation process does not change the magnitude of the field representation but redistributes it among the sensor coils. This stands in contrast to the position solution which directly controls the magnitude of the signal received from the three source coils.

The position solver operates on the un-rotated signal matrices when it matches (3.5) to find the position estimate (3.23). Position errors are indicated when there is a difference between the un-rotated version of the estimated and measured signal matrices, leading to a non-zero result of the minimization equation. Although these differences are clearly related to the position error, they are obscured by the multiplications in (3.5). A better measure of the error in the position estimate is the difference between the estimated and measured signal matrices without rotation. Unfortunately, the un-rotated measured signal matrix is not available but it can be approximated. Using the estimated sensor orientation (q), the measured signal matrix (S) is rotated into the source reference frame and then subtracted from the un-rotated signal matrix estimate ($f_D(r) + g((r))$) to obtain the error in the signal matrix (S) of the estimated position, referred to as the signal matrix estimation error (A) (3.30).

$$\Lambda\left(r, q, S\right) = \left(f_D\left(r\right) + g\left(r\right) - QtoT\left(q\right) \cdot S\right) \tag{3.30}$$

The estimation error (Λ) is used to develop a polynomial-based estimate of the sensor PnO errors with multivariate regression. Once the error in the sensor 0 PnO is known, the error at each sensor can be estimated through the known relative PnO. Modifying (3.18) to include error factors shows that the sensor 0 position estimation (r^{erl}) adds an offset to the mapping fixture position estimate while the orientation error (T_k^{err}) adds additional rotation (3.31).

$$r_k^n = \left(r_k^0 + r^{err}\right) + \left(T_k^{err} T_k^M\right)\left(rP_0^n\right) \tag{3.31}$$

Expanding (3.31) an equation for the position error of sensor n at time step k (ϵ_k^n) is developed as a function of the sensor 0 position error (ξ_k^r) and sensor 0 orientation error (ξ_k^{AER}; expressed in Euler angles) (3.32). The orientation error at sensor 0 adds rotation to the fixture sensor orientation through the fixture frame (3.33).

$$\epsilon_k^n = \xi_k^r + \|rP^n\| \tan\left(\|\xi_k^{AER}\|\right) \tag{3.32}$$

$$q_k^n = q^{err} \otimes \left(q^M \otimes qP^n\right) \tag{3.33}$$

The estimated sensor error is used to generate mixing coefficients for the data interpolation process.

3.4.3   FIELD DATA COLLECTION

The data collection process uses a six-step procedure to assemble the data to create a map of the secondary field using the interpolation fixture (Fig. 3.5) and the mapping fixture (Fig. 3.6).

Step 1 installs eight sensors on the interpolation fixture to form a virtual cube with a sensor on each vertex. Step 2 places the fixture in a non-distorting environment and the calibration data consisting of at least 100 simultaneous measurements of the sensor PnO. After completing the calibration process, step 3 moves the interpolation fixture to the center of the mapped volume to define the interpolation volume. A large data set of magnetic field measurements are made with the fixture to determine the secondary field value used in the interpolation process. Step 4 moves the sensors from the interpolation fixture to the mapping fixture. Step 5 places the mapping fixture in a non-distorting environment to collect calibration data and determine the sensor PnO in the fixture reference frame. Finally, Step 6 collects a large data set of field measurement by slowly moving the mapping fixture through the mapped volume while simultaneously collecting field data from all fixture sensors.

Step 1: Interpolation Fixture Assembly   The first step in the data collection process installs the eight position sensors on the vertices of the interpolation fixture. The sensor configuration on the fixture defines the fixture reference frame through r^A and r^B (Fig. 3.2).

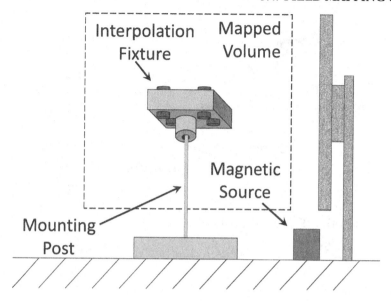

Figure 3.5: The interpolation fixture is placed on a mounting post designed to center it in the mapped volume. The fixture is aligned with the source reference frame and then field data collected to determine the secondary field values.

Step 2: Interpolation Fixture Calibration Data   Once the interpolation fixture is assembled it is moved to a non-distorting environment for the calibration process (step 2). The fixture is place at a reasonable distance to the magi entice source (~0.5 m) and a large data set of PnO measurements is taken. There is no requirement to locate the fixture with a specific pose for the calibration process but it should be stationary to provide the highest degree of accuracy. Once the calibration data is captured, care must be taken to avoid moving the sensors from their calibrated positions until they are moved to the mapping fixture.

Step 3: Interpolation Fixture Field Data   Step 3 of the data collection process collects field data with the interpolation fixture to determine the secondary field at the fixture sensor positions (Fig. 3.5.). The interpolation volume must be located so that the mapping fixture can keep one sensor inside the cube while reaching all areas of the mapped volume. Additionally, the PnO of the fixture must be determined to declare a known PnO that is reasonably close the actual value. The tracker cannot be used for this measurement since the distortion in the environment will cause incorrect readings, so mechanical methods must be used. The declared PnO of the fixture should be a close approximation of fixture PnO (Fig. 3.2.).

Once the interpolation volume has been defined, the fixture is placed in the appropriate position and aligned with the source reference frame at the known orientation. With the inter-

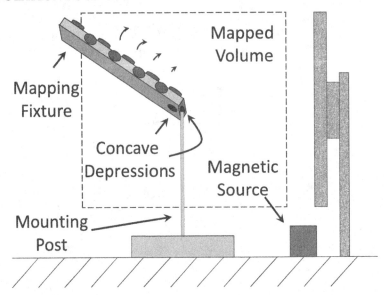

Figure 3.6: The mapping fixture was designed to work with the same mounting post used with the interpolation fixture. The fixture has a concave impression that allows the fixture to be rotated about the mapped volume while keeping sensor 0 inside the interpolation volume.

polation fixture now placed in the mapped volume, 100 points or more of field data is captured (see Section 3.4.4) and used to determine the secondary field at the cube vertices. The fixture is now removed from the environment and the sensors moved to the mapping fixture.

Step 4: Mapping Fixture Assembly The first three steps have used the interpolation fixture to construct an interpolation volume in the map with known field values. In step 4, the sensors are mounted on the mapping fixture in preparation for collecting field data for the map. The placement of the sensors on the fixture can theoretically be arbitrary but it should have a maximum 38.1 mm offset between each consecutive pair. This spacing requirement matches the grid offset used with standard mapping procedures and has been shown to produce accurate interpolation results.

Step 5: Mapping Fixture Calibration Data The mapping fixture simultaneously collects field data at each of the sensor locations while using the interpolation volume to estimate the PnO of each sensor. Before data collection can begin the fixture must be calibrated in a non-distorting environment to determine the PnO of each sensor in the fixture reference frame. A large dataset (> 100) of simultaneous PnO measurements is collected for the mapping fixture sensors. After the calibration data collection the equipment is moved back to the mapped volume.

Step 6: Map Field Data Collection   The final step of the process uses the mapping fixture to collect a "cloud" of field points for the map generation process. To collect data, the fixture is positioned so that sensor 0 is inside the interpolation volume while the fixture is slowly moved through the mapped volume (Fig. 3.6.). During this time, the tracker is continuously collecting field data for each sensor. The data collection process continues until a dense data cloud of points covering the entire mapped volume has been collected.

3.4.4   LOOK-UP-TABLE (LUT) GENERATION

The collected data consists of trajectories through the tracking volume that create a large set of scattered points, each with an associated magnetic field measurement. To create the map, the PnO of each data point must be determined to normalize the data to the source reference frame and associate it with the correct LUT grid positions. The map creation process consists of six distinct (Fig. 3.7), (Step 1) interpolation fixture calibration, (Step 2) interpolation volume parameterization, (Step 3) mapping fixture calibration, (Step 4) sensor PnO estimation, (Step 5) data tabulation and finally, (Step 6) secondary field data interpolation.

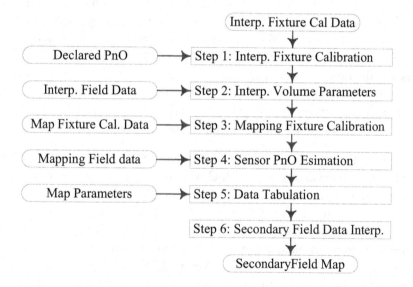

Figure 3.7: The Look-up-table (LUT) is created from collected data through a six-step process.

Step 1: Interpolation Fixture Calibration   The map generation process begins with the calibration of the interpolation fixture. The calibration is actually a software task and is performed after the data collection is complete but it is dependent on the correct placement of the sensors on the interpolation fixture. The interpolation fixture is assembled by placing the eight sensors in positions associated with their XYZ coordinates in the fixture frame (Fig. 3.2.). Once the sensors are installed in the correct configuration a set of calibration data is captured in a non-distorting environment. The calibration data is averaged over the data set to remove measurement noise before constructing the fixture reference. The sensor orientation is averaged as Euler angles and then converted to quaternions. The relative sensor positions in the fixture frame (r^P) can then be calculated using (3.14) and (3.15).

Step 2: Interpolation Volume Parameterization   To parameterize the interpolation cube, the declared interpolation fixture PnO is combined with the magnetic field measurement data to initialize the interpolation volume parameters. The user declares the fixture PnO in the mapped volume by specifying the PnO of interpolation fixture sensor 0. The PnO of the other fixture sensors in the mapped volume is "backed out" of the declared sensor 0 PnO using the relative PnO (r^P, q^P) derived from the calibration data. The dipole field at each of the cube vertices is now calculated using the dipole function (3.3). The secondary field at the vertices is the difference between the average measurements and the ideal dipole values.

Step 3: Mapping Fixture Calibration   The mapping fixture uses multiple sensors in a fixed geometry that must be characterized to determine the relative PnO of the sensors on the fixture. A reference frame is defined for the mapping fixture using the process preciously described in Section 3.4.4. The calibration position data is averaged and then used to determine the sensor positions in the fixture reference frame (r^B) using (3.25). The orientation data is averaged as Euler angles and then converted to quaternions before calculating each sensor orientation in the fixture reference frame with (3.26). Unlike the interpolation fixture, there is no requirement for precise placement of sensors on the mapping fixture; therefore, no optimization of their locations is required. The calibration data is simply averaged and a referenced frame created.

Step 4: Sensor PnO Estimation   Once the relative sensor PnO on the mapping fixture is known, the sensor 0 PnO is used to determine the PnO of all the other sensors on the fixture. The sensor 0 PnO is estimated using the optimization process described in Section 3.4.4. The PnO of sensor 0 is estimated for each data frame from the associated field data, and any points outside the interpolation volume are discarded to limit the estimation errors. The points inside the interpolation volume are used to estimate the PnO of all the fixture sensors from the relative PnO. The fixture reference frame origin is at sensor 0; therefore the fixture position is known. The position of fixture sensor n at time k is the sum of the sensor 0 position at time k ($r_0(k)$) and the relative position of sensor n on the fixture (r_n^B), rotated by the mapping fixture orientation in the source reference frame ($T^M(k)$) (3.34). The rotation matrix representing the mapping fixture orientation is determined by the sensor 0 orientation (3.27).

$$r_n(k) = \left(r_0(k) + T^M(k) \cdot r_n^B\right) \tag{3.34}$$

The sensor orientation at time k is calculated as the quaternion product of the sensor orientation in the fixture frame (q_n^B) and the sensor 0 orientation at time k ($q_k(k)$) (3.35).

$$q_k^n = q_n^B \otimes q_0(k) \tag{3.35}$$

Step 5: Data Tabulation Each sensor on the mapping fixture is potentially at a different but known orientation. To combine the field data into a single table the measured field data must be rotated into alignment with the source reference frame. The secondary field is found by rotating the measured signal matrix into alignment with the source reference frame and subtracting the ideal dipole field of the point (3.20). The estimated PnO and calculated secondary field values are combined into a table with one row for each data point. This table is now indexed to the output map format using map parameters supplied by the user. The required parameters specify the minimum grid point position in the source frame, the grid increment and the number of grid points for each axis. The three parameters are used to determine how the interpolation volume fits into the map volume and to index the data. For example, a map with the number of grid points equal to [4, 4, 4] and an increment of 0.0381 mm would have 64 cells in it, each cell being a cube with a dimension of 38.1 mm.

The data collected by the mapping fixture forms a sphere centered at the interpolation cube while the generated map will be rectangular, leaving a large quantity of collected data points outside of the mapped volume. To include these points in the interpolation process the size of the mapped volume is increased by one cell on each surface of the rectangular mapped volume. The additional cells provide the data needed to interpolate the grid points at the edges of the map.

The collected data is now indexed to the map to provide a method of sorting and searching the dataset. The index value is computed by assuming the table is sorted in XYZ format, with X as the MSB and Z as the LSB. The index (Δ) is computed from the grid coordinate (α) and the number of grids per axis (γ) (3.36). The grid coordinate refers to the grids point index while the grid per axis is the number of grids per axis, each organized in XYZ vector format.

$$\Delta(\alpha, \gamma) = i\alpha_0 \cdot \gamma_1 \cdot \gamma_2 + \alpha_1 \cdot \gamma_2 + \alpha_2 \tag{3.36}$$

The data table is now placed into XYZ order by sorting on ascending index. This data ordering is used to minimize the number of iterations required to assemble localized data for the grid point interpolation process.

Step 6: Secondary Field Data Interpolation The interpolation process uses the data in the surrounding eight cells to estimate the secondary field at each grid point. The data required for each grid point interpolation is assembled by generating the index of each of the surrounding cells, sorting it into ascending order and then searching the table for matching indices. The ascending order of the table allows us to assemble all the local data values for a grid point with one pass through the table.

To interpolate the data, a multivariate polynomial regression is used to fit a function to each element of the secondary field in the locality of the grid point from a dataset including all points within a 0.0381 mm range. The position of each data point (r) is used with a polynomial term generation function ($f_p(r)$) and the associated collected data ($S(r)$) to determine the polynomial coefficients (3.37). Once the coefficients are known, the signal matrix at the grid point ($G(r_{gp})$) is estimated using the polynomial function with the position grid point ($X(r_{gp})$) as the argument (3.38).

$$C = (XX^T) \cdot X \cdot S(r) \quad X = f_P(r) \tag{3.37}$$

$$G(r_{gp}) = X(r_{gp})^T C \tag{3.38}$$

The interpolation process generates a single secondary field vector for each grid point in the user-defined map.

3.5   EXPERIMENTAL RESULTS

A tracking volume with moderate distortion was used to test the mapping procedure and fixtures. The map was defined as the area directly in front of a large LCD screen, with the LCD being perpendicular to the X-Y plane of the map (Fig. 3.8).

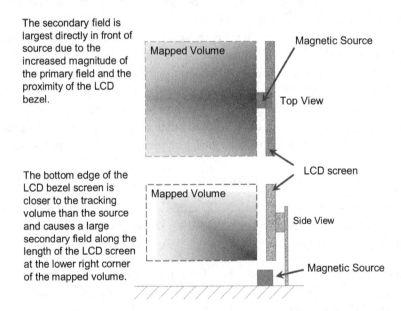

Figure 3.8: The secondary field shown as a percentage of measured signal for our experimental data. The darker regions indicate areas where the secondary field is a largest, illustrating how source location directly impacts the distortion level.

A Polhemus mapper was used to generate a map of the secondary field with known position and orientation in the source reference frame. This map was used to determine the secondary field at arbitrary field points in the mapped volume. The Polhemus mapper by design has an absolute PnO error less than the tracker (Table 3.1) and provided accurate magnetic field measurements on a uniform 38.1 mm grid.

Table 3.1: IVC: PnO measurement equipment accuracy specification

Equipment	Static position Error (mm)	Static orientation Error (milliradians)
Mapper	< 0.25	< 1.4
Liberty tracker	0.71 (RMS)	2.6 (RMS)

Tri-linear interpolation was used to generate the field point data from the secondary field table. The measured field at any field point in the mapped volume can be found as the sum of the estimated secondary field and the dipole Equation (3.3). This method was used to generate the measured field for test data from PnO measurements taken in the same volume without distortion (the LCD screen was removed).

3.5.1   PNO ESTIMATION USING INTERPOLATION

The accuracy of the PnO estimates is dependent on the ability of the cost function minimization process to estimate the PnO in the interpolation volume. To evaluate the performance of the tri-linear interpolation process in this application, a trial data collection was simulated using a virtual interpolation cube placed in the center of the mapped volume. The analysis used the mapping fixture to collect 100,000 samples of sensor 0 PnO data as the fixture was moved about the mapped volume collecting field data.

The measured field of each point was estimated as the sum of dipole field ($f_D(r)$) and an interpolated secondary field (G) at the field point (r), rotated by the measured sensor orientation (3.4). The secondary field was interpolated from the LUT constructed with the Polhemus mapper data. A position solution was calculated for each field point using the position cost function minimization and then compared to the known field point position for error analysis. The cost function was recursively seeded similarly to what occurs in the mapping process. Similarly, the orientation was solved using the cost function minimization.

The interpolation function performed very well, giving a median position error of 0.27 mm and median orientation error of 1.50 milliradians (Table 3.2). Looking at the histogram for the position error (Fig. 3.9(a)) we see that the majority of the points have very small errors but there is a long "tail" of large error values. Although most of the interpolation position errors were within the tracker error band (0.71 mm in Table 3.1); there is a large group of outliers (position errors > 0.91 mm) with errors ranging as high as 4.9 mm.

The orientation error (Fig. 3.9) has a less focused distribution and the majority of the points are larger than the maximum orientation error specified for the tracker (0.41 milliradians; Table 3.1). The increased error in the orientation estimate is indicative of its dependence on the

Table 3.2: IVC: PnO error profile

Measurement	Position (mm)	Orientation (milliradians)
median	0.27	1.50
75 percentile	0.50	2.40
Max	4.9	23.4

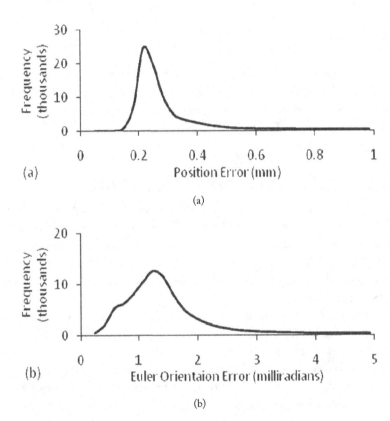

(a)

(b)

Figure 3.9: Histograms of the position error (a) and orientation error (b) in Interpolation estimates have a long tail of outlier values.

accuracy of the sensor position estimate. The orientation cost function (3.22) minimizes the error between the signal matrix at the estimated position and the measured signal matrix. When the position error is small it has little effect on the orientation estimate but when the error is large, the orientation estimate error increases dramatically. This relationship results in error accumulation in the orientation estimate, with a broader distribution and more outliers than the position error.

Both distributions had a large number of outlier values (17% for position errors; 16% for orientation) indicating that the errors may be the results of a mixture of distributions. An examination of the data shows that the errors are related to at least two sources, the measurement itself (of the field data) and the inability of the cost function solver to converge on the correct answer at some field points. The larger errors are grouped in long smooth trajectories of sequential points with errors of 1 mm or more. The positions suggest that the PnO solver is having difficulty converging on the correct answer. These paths represent groups of solutions that diverge slightly from the true result due to convergence in local minima and are most likely related to our use of recursive seeding. This method helps the solver remain converged on the true path of the sensor by starting the solution search very close to the correct answer.

3.5.2   MAPPING FIXTURE ACCURACY

To determine the errors at each sensor on the mapping fixture we estimate the sensor 0 PnO error from the error in the estimated signal matrix using Λ (3.30). A strong correlation between the error in the un-rotated signal matrix estimate of a solution (Λ) and the position error is expected since the range solution is directly related to the magnitude of the signal matrix.

As shown in Fig. 3.10, the RMS of Λ has a strong correlation to the position error and can provide the basis for a correction of the position estimate. The orientation estimate also has a strong correlation to the signal matrix estimate that is well defined (Fig. 3.10) but is "looser" than that of the position error. Closer examination reveals that the majority of the points with large errors are grouped together and can be separated from the majority of the data. Imposing a maximum value of 0.3 on Λ for all data points removes most of the large errors from the sample population and improves the ability of the polynomial-based error estimator to accurately predict the interpolation errors.

The strong correlation between the PnO error and the Λ metric suggests that a polynomial with the elements of Λ is a robust method of estimating the interpolation error. Unfortunately the interpolation errors are in close proximity to the tracker error band (Table 3.1) and the polynomial-based error estimate was sensitive to noise in the measured signal matrix, resulting in poor estimates. Based on speculation that the interpolation PnO errors were grouped in specific regions, a new polynomial estimator was constructed using the estimated position (r) and the signal matrix error (Λ) as independent variables.

To evaluate the performance of the estimator, the difference between the estimated and measured error for a 100,000 point population was used, assigning a negative magnitude if the measured error was larger than the estimate (Table 3.3). In the following discussion of the results the error is computed as the difference between the estimated error and the actual error. A case where the actual error is larger than the estimate will be assigned a negative value and labeled "undershoot." When the estimate is larger than the actual error it will be labeled an overshoot. Overshoot conditions are of no concern since they reduce the effect of the associated sample when the weighting is applied and have little effect unless they occur at a high rate. Undershoot

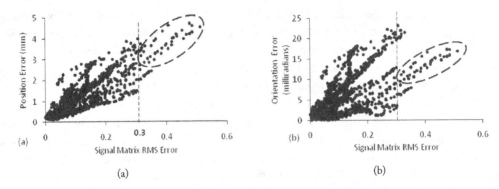

(a)

(b)

Figure 3.10: The PnO error of the interpolation estimate has a correlation to the RMS error of the unrotated signal matrix estimate. In this figure we plot the position (a) and orientation (b) errors against the RMS average of the error in the un-rotated signal matrix. The data appears in long strings of closely placed errors due to the combination of a high measurement rate (240 Hz) and slow motion of the fixture. Note the cluster of large errors at the upper right corner of both plots; these groups of outliers are removed by imposing a maximum RMS error constraint 0f 0.3 on the estimated signal matrix.

conditions allow sample errors to be weighted much greater than appropriate and can skew the LUT interpolation if the associated error is large.

Table 3.3: IVC: Error estimation "goodness of fit"

Measurement	Position (mm)	Orientation (milliradians)
Max. Undershoot	-3.90	-17.75
5 percentile	-0.39	-2.27
25 percentile	-0.05	-0.55
50 percentile	0.00	-0.22
75 percentile	0.03	0.00
95 percentile	0.59	2.60
Max.	86.5	412

The position estimates were within 0.6 mm for 90% of the samples while the orientation was within 2.6 milliradians for the same range. The position estimator undershot by more than 1 mm in 5% of the samples but was within 50% of the measured value in all cases. The orientation also had approximately 5% of the samples with undershoot but again the estimate was within 50% of the actual value. Both estimators had large undershoots for 5% of the sample, effectively removing those points from the experiment. In this case, the overshoots are spread about the volume and they only represent 5% of the population so they can be ignored.

In the experiment the estimated error was used to establish a maximum allowable error for measurement error. The total position error in a field measurement was computed using (3.24) to estimate the error at each of the sensors on the mapping fixture. This threshold was implemented in the map-generate process to exclude data with an estimated position error greater than 1 mm (Table 3.4). A 1 mm boundary will exclude approximately 10% of the collected samples, including the majority of the undershoot/overshoot conditions.

3.5.3   FIELD DATA COLLECTION

The data collection process uses the two interpolation fixtures to collect magnetic field data in the mapped volume. The two fixtures were built from wood stock and then used with the procedure detailed in Section 3.4.3 to collect data.

The interpolation fixture was constructed with precisely placed mounting holes for each sensor to align them with the vertices of the interpolation volume. The sensors however are not manufactured with a precision-mounting surface resulting in average position error of 1.26 mm (Table 3.4).

Table 3.4: IVC: Interpolation fixture avg. sensor position errors (MM)

Sensor	0	1	2	3	4	5	6	7
X	0.0	-0.9	1.3	1.6	-1.2	0.0	0.9	1.7
Y	0.0	-2.2	-0.8	-1.4	0.0	-2.3	-2.9	-2.1
Z	0.0	-0.3	0.0	-3.3	0.0	-3.3	0.4	-3.3

To improve the accuracy of the signal matrix interpolation, the interpolation cube size and rotation were optimized to best fit the measured data (3.37). This optimization function varies the cube side dimension (d), cube offset (r) and orientation (in the fixture reference frame) (q_C) to minimize the errors at each cube vertex using an LSS minimization process. Note that a standard conversion from quaternion orientation to rotation matrix is performed in (3.39). The cube orientation (q_C) is optimized in quaternions and then converted to a rotation matrix (T_C) using a standard conversion function.

$$\arg\min\left[rP - QtoT\left(q_C\right)\Psi\left(r, d\right)\right] \tag{3.39}$$

$$\Psi\left(r, d\right) = r + d \begin{bmatrix} 0 & 0 & 0 & 0 & 1 & 1 & 1 & 1 \\ 0 & 0 & 1 & 1 & 0 & 0 & 1 & 1 \\ 0 & 1 & 0 & 1 & 0 & 1 & 0 & 1 \end{bmatrix}$$

Using the optimized cube as a new fixture reference frame, the previously computed relative sensor positions (r^P) were translated into that frame. The relative sensor orientations (q^P) were calculated using the new fixture frame position in the source frame and averaged quaternion measurements. The interpolation cube optimization process removed most of the sensor placement errors, leaving a few offsets in the 1 to 2 mm range (mm) (Table 3.5). The remaining errors did not have a large effect on the interpolation accuracy due to the small magnitude of the errors as compared to the interpolation volume dimensions.

Table 3.5: IVC: Optimized interpolation fixture sensor position errors (MM)

Sensor	0	1	2	3	4	5	6	7
X	0.4	-0.9	0.3	0.2	-0.5	-0.2	0.2	0.6
Y	0.3	-1.7	0.2	-0.2	1.6	-0.4	-0.6	0.5
Z	-0.2	2.1	-0.4	-1.1	0.2	-0.4	0.5	-0.6

To gauge the importance of the sensor positions on the interpolation cube, a statistical analysis of position errors at random points in the interpolation volume was run. The test was run for 10,000 points of data captured in a non-distorting environment while simulating data collection. Three interpolation cubes were considered, an "ideal" cube with no position errors, the "optimized" cube previously discussed and a cube constructed from the "averaged" data without optimization.

Looking at errors at the farthest sensor from the interpolation volume (Fig. 3.11) we see that the ideal and optimized volumes reduce the median error but do not have the same impact on the larger errors (0).

Table 3.6: IVC: Sensor 8 position error vs. interpolation cube

Measurement	Ideal	Optimized	Averaged
Median	0.48	0.68	4.31
95 percentile	9.54	12.0	12.69
Max.	16.8	16.4	16.7

The ideal volume does reduce the 95% confidence interval error by approximately 25% but the optimization has almost no effect on the larger errors, suggesting that these errors are not related to the sensor positions on the fixture. These larger errors most likely correspond to specific areas in the interpolation volume where the tri-linear interpolation process has difficulty accurately estimating the secondary field data. The "trouble spots" may be localities of the interpolation volume that have a high degree of non-linearity in the field data and are not well estimated by a linear function. The interpolation estimates are critical to the accuracy of the field measurements made with the mapping fixture and must be handled correctly to avoid large errors entering the LUT we are creating. Errors in the orientation estimates will have a larger effect on the accuracy of the LUT we are building since they are multiplied by the displacement of sensors on the mapping fixture from sensor 0 (the one inside the interpolation volume). The impact of these errors is reduced when they are weighted with the error estimation in the LUT generation process.

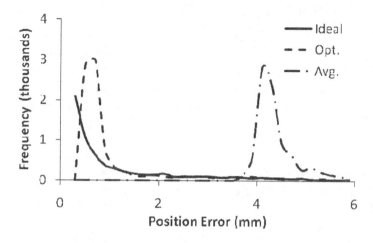

Figure 3.11: The lack of a precision-mounting surface on the sensors results in positioning errors on the interpolation fixture. Here we see the position error distribution of mapping fixture sensor 8 (the farthest away from the interpolation volume) when using each of the three interpolation volumes (ideal, optimized and averaged) to estimate sensor positions. Note the large reduction in error when using the optimized cube as compared to the measured one.

3.5.4 LOOK-UP-TABLE (LUT) GENERATION

An LUT of secondary field measurements was created from the collected field data using the process outlined in Section 3.4.4. To evaluate the accuracy of our LUT, two PnO generation tests were run; the first compared the grid points of the Polhemus LUT to one generated with IVC while the second used the field map to correct PnO estimates and then compared them to the dipole (uncompensated) solution. In both these experiments, secondary field data from the IVC LUT is used to correct the field measurement before solving for PnO.

The comparison of the on-grid secondary field estimates of the Polhemus map and the IVC map was conducted by solving for the PnO at each grid point. The field data for each point was generated from the Polhemus map to provide a benchmark to compare our map against. The PnO solution was found by subtracting the secondary field from the field measurement and then solving using the dipole algorithm.

The IVC map had errors of up 48.7 mm with a median of 24.7 mm and the orientation error had a median of 75 milliradians with max of 280 milliradians (Table 3.7).

The position error distribution is very narrow while the orientation error distribution is somewhat broader. These errors are larger than expected given that the error in the secondary field estimate is typically less than 5% of the RMS sum of the estimate (Fig. 3.12). The mismatch of the lower magnitude elements causes small errors in the signal matrix that skew the range solution (3.6) slightly and have a cascading effect on the position (3.6) and orientation (3.7) solutions.

Table 3.7: IVC: On-grid PnO error for ICV LUT

Measurement	Position (mm)	Orientation (milliradians)
25 percentile	19.6	49.9
50 percentile	24.7	74.7
75 percentile	29.2	102
99 percentile	40.2	191
Max.	48.7	280

The orientation solution is impacted more by these small errors since it is dependent on the ratio metric relationship between the nine elements of the signal matrix.

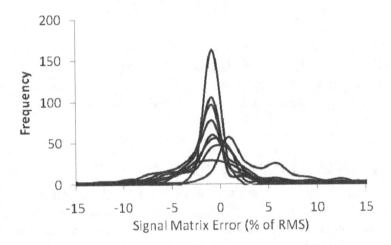

Figure 3.12: A histogram of the error in the individual elements of the secondary field estimates shows that the majority of the errors are well below 5% of the RMS value of the signal matrix.

A comparison of the compensation provided by the LUT map was conducted by randomly choosing 10,000 points with random rotation in the mapped volume. Each point was solved with secondary field compensation and without to illustrate the ability of the LUT to correct PnO measurements in a distorting environment. The PnO solution provided by the IVC system closely tracked the results we saw with for the grid point comparison.

The position error was in a tightly grouped band with a median of 28.1 mm while the orientation had a broader distribution with mean of 38 milliradians (Table 3.8). Note that the position error has essentially the same distribution as the grid comparison but with a much larger maximum error (240 mm). The largest errors (outliers) are most likely caused by a firmware problem with the solver since this large an error indicates that the solver converged on a solution outside of the mapped volume.

Table 3.8: IVC: Comparison of PnO error

Measurement	Position (mm)		Orientation (milliradians)	
	IVC	Dipole	IVC	Dipole
25 percentile	25.0	79.2	22.7	1853
50 percentile	28.1	113	38.0	2598
75 percentile	30.9	150	60.0	3248
95 percentile	34.9	186	106	3744
99 percentile	38.4	204	156	4128
Max.	240	333	887	4471

Comparing the position solution to the non-compensated case (Fig. 3.13), the IVC system provides better than a 75% improvement in the position solution. The compensated solution had a sharp peak in the error distribution, with all solutions (except the maximum) within 50 mm of the correct solution. This compares very favorably with the essentially uniform distribution of position errors for the non-compensated case.

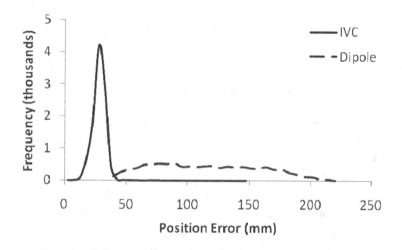

Figure 3.13: A chart of the histogram of position error for the IVC corrected data and uncorrected (dipole) data. The IVC errors are clustered below 50 mm as compared to the broad, almost uniform distribution of the dipole function data.

The orientation error has a broader distribution (Fig. 3.14) than position but again the majority of the error is small (less than 50 milliradians). Note that the random point test resulted in better orientation estimates than grid point test. This improvement is attributed to the effect of interpolating the LUT secondary field estimates. Errors in a grid point estimate will dominate the interpolated secondary field estimates when taken at that location in the map because the interpolation is a "pass-through" function in this situation. Taking random points in the map, all eight grid point estimates surrounding the field point are used to estimate the secondary field. An error in a grid point estimate is somewhat mitigated by the other values used in the interpolation process. Errors in more than one of the eight values used in the interpolation will not accumulate since they are unlikely to occur in the same element of different grid point estimates.

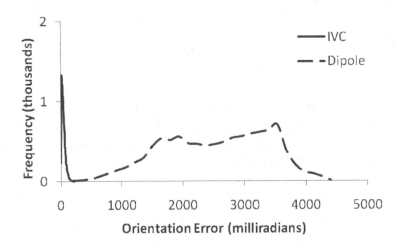

Figure 3.14: A histogram of the orientation error for the IVC and dipole PnO system illustrates the large improvement in orientation accuracy with our system.

The IVC system orientation error was greatly improved as compared to the uncompensated case. Looking at the orientation error (Table 3.8), the IVC map has less error than the dipole case. Orientation errors of up to 4.47 radians occur for the uncompensated case as compared to a maximum error of 0.887 radians for the IVC estimate. If we ignore the outliers and look at the 99 percentile, we see that the IVC system is within 156 milliradians of the correct solution. The improved performance of the IVC system is also evident in the relatively compact orientation distribution and compared to the uncompensated case. The nearly uniform distribution of the uncompensated case along with the very large error values suggests that these measurements are essentially unusable without compensation.

3.6   SUMMARY

The IVC system provides a low-cost, accurate method of generating LUT-based field compensation for AC electromagnetic trackers. The system provides substantial improvements in tracker accuracy when used in moderate to highly distorted environments. Test data shows that the uncompensated tracker PnO is essentially unusable without the IVC field correction due to gross inaccuracies. Specifically, the orientation error was as large as 4 radians in some cases and the error distribution has a wide range (Fig. 3.13). With a 3σ confidence region, the orientation error of the IVC system drops to 0.156 radians with a median of 0.036 radians, a more than 10x improvement over the uncompensated case.

Some areas of potential improvement for the system include the following.

- The lack of a precise mounting surface on the sensors created a problem in locating the sensors on the interpolation fixture. This problem was addressed with an optimization process to minimize the location error but it was unable to correct all of the errors. The accuracy of sensor PnO estimation in the interpolation volume is dependent on each fixture sensor being correctly located. This problem could be eliminated using a precision fixture in conjunction with sensors that have been modified to present a precise mounting interface.

- Levenberg-Marquardt minimization (LMM) was used for both the position and orientation cost functions without constraints and this is an example of the shortcomings of this approach. The minimization has no information on the range of the solution (based on previous estimates) with which to reject unreasonable answers. Generally, this was not a problem since the position errors were small, but the outlier errors caused correspondingly large orientation errors. This was also a problem in our LUT-based compensation where an LMM cost function was used to optimize the solution. A Kalman filter-based approach might be better suited to these tasks.

- The polynomial interpolation used to generate the on-grid estimates had difficulty accurately estimating the smaller elements of the secondary field signal matrix. This result is related to the spatial distribution of the collected data and the use of an LSS type estimator. Due to the manner of the data collection, the field data is located in concentric spheres centered at the origin. These spheres had an offset of 25.4 mm, meaning that a typical grid data point was interpolated from two bands of data that pass through the 76.2 mm diameter interpolation space. The interpolation result could be improved if a more varied data set was created through more closely spaced sensors. Additionally, a different estimator such as nearest neighbor interpolation could be chosen to improve the quality of the interpolation itself.

The ICS system answers a need for a low-cost field mapping system for applications that require moderate accuracy. The system reduces cost in two ways: it can be constructed by the user from widely available, inexpensive materials, and it does not require highly trained individuals to operate. The construction cost of the fixtures used in this experiment was literally less than $20.00. Improved results could be achieved by spending a few hundred dollars on an improved interpolation fixture. The data collection process takes approximately one or two hours, including time for gathering calibration data for the two fixtures and location of the mapping pole in the target environment.

CHAPTER 4

Conclusion

This project has addressed the difficulties that occur when trying to use an AC electromagnetic tracker in a simulation environment with head tracking. The head tracker is used to provide information on the user's line of sight and to predict head motion for display lag compensation. A novel Extended Kalman Filter, the DQEKF, was developed to address the need for an efficient predictor of quaternion head orientation. A new method of creating magnetic field maps for AC magnetic trackers was developed and presented. The new mapping technique provided significant improvement in tracker performance without the use of precision measuring equipment.

The DQEKF development was successful in developing a small, efficient Extended Kalman Filter for head orientation prediction. This method uses a three-step framework to provide a computationally efficient mechanism for predicting future orientation within 4D quaternion space. The DQ framework was compared to the quaternion EKF (Q) filter using head motion data representing three individual categories of head motion with prediction intervals varying from 0 ms to 116 ms. Additional experiments were conducted with data sets representative of head motion in a VR/AR environment. Experimental results show that the DQ approach provides prediction performance similar to quaternion EKF while requiring only a fraction of the computational load.

The delta quaternion filter was expanded into the multiple model delta quaternion filter to deal with aggressive head motion. The MMDQ provides a natural extension to the DQEKF development and there has been no published work using a DQ filter for prediction in a multiple model framework. The MMDQ2 (the two-filter version) provided excellent prediction performance, matching the DQEKF-CA during low accelerations and greatly improving on performance for aggressive motion.

This project also targets the ability to use AC electromagnetic trackers in environments with conductive or ferrous materials. Traditionally these environments have very difficult to work in, requiring careful control of materials used in the simulator. In many cases field mapping is required to provide the level of performance required by the application. Currently there is no method of constructing a field map for this kind of system without contracting the manufacturer. The complexity and expense of collecting field data to create an LUT for field corrections requires highly trained individuals with specialized high precision equipment. While it is true that many applications require the high accuracy that this kind of mapping provides, most only require a moderate level and do not warrant the time and expense; the IVS system was specifically designed for these applications. The system uses easily constructed fixtures that do not require high precision machining. The data collection process is easily understood and does not require special training to obtain excellent results. Although the system software was implemented in Matlab for this experimental work, implementation as a customer-operated utility does not require extensive investment. Results obtained with the IVC system were excellent, allowing operation of a Polhemus Liberty tracker in an environment that had a demonstrated need of field mapping. Although the IVC results do not equal those of the Polhemus mapping system in terms of compensated accuracy, the low investment of equipment and manpower required by IVC are more than offset by the large improvement in tracker performance that was obtained. The LUT generated for the LCD test case reduced orientation errors by more than an order of magnitude when compared to the uncompensated tracker.

This project has resulted in three major contributions:

1. Development of an efficient head orientation prediction methodology (DQEKF) suitable for small real-time systems such as electromagnetic trackers.

2. Extension of the DQEKF concept to multiple-model filtering that is able to accurately predict head motion for the full range of head motion, including aggressive motion.

3. Invention of a new approach to electromagnetic field mapping for AC electromagnetic tracker applications.

The result of this work is that an AC electromagnetic tracker will be usable in most environments without the expense of high precision mapping. This technology is directly applicable to other areas besides simulation in the medical, consumer and military fields.

APPENDIX A

The Delta Quaternion Extended Kalman Filter (DQEKF)

The Matlab code used for each segment of the paper (DQEKF, MMDQ and the IVC) are presented below. Common functions not available in the Matlab library are included in Appendix D.

A.1   RUNDQEKF

The function runs the DQEKF on a sequence of orientation measurements. This version supports multiple prediction models including constant velocity (CV), constant acceleration (CA), and an alternate constant acceleration model (CA2).

Input parameters:

qIn	orientation measurements
Q	the process noise
R	the measurement noise
t	the time step (seconds)
m	model ID (CV: constant velocity; CA: constant acceleration, CA2: alternate CA)

```
function  dOut=RunDQEKF(qIn,Q,R,t,m,f)
%a function to execute the DQ EKF on quaternion input data
% qIn    input data arranged as a series of quaternion vectors
% Q      the process noise covariance matrix
% R      the measurement noise covariance matrix
% t      time base in seconds
% m      model used (CV:constant velocity, CA: constant acceleration)
% f      debug flag

N=length(qIn(1,:)');

CV=0;
CA=1;

Acv=[[eye(3,3),eye(3,3)*0];[eye(3,3)*0,eye(3,3)*0]];
Aca=[[eye(3,3),eye(3,3)*t];[eye(3,3)*0,eye(3,3)]];
```

```
Wcv=[eye(3,3)*t;eye(3,3)*0];
Wca=[eye(3,3)*t*t/2;eye(3,3)*t];

%get A and W
if (m==CV)
    A=Acv;
    W=Wcv;
else
    A=Aca;
    W=Wca;
end
WQW=W*Q*W';

if (f==1)
    R=R(2:4,2:4);
end

dq=zeros(4,N);
for k=1:N-1
    dq(:,k)=Qmul(qIn(:,k+1),qIn(:,k),0,1);
end
dq(:,N)=dq(:,N-1);
if (f==0)
    zm=dq;
else
    zm=dq(2:4,:);
end

x=[1e-12,1e-12,1e-12,1e-12,1e-12,1e-12]';
if (f==0)
    xOut=zeros(14,N);
else
    xOut=zeros(12,N);
end
I=eye(6,6);
P=eye(6,6);

for k=1:N
```

```
        x=fMM(x,t,m);
        ze=GetZ(x,t,f);
        H=GetHmmR(ze,x,t,m,f);
        P=A*P*A'+WQW;
        v=zm(:,k)-ze;
        S=H*P*H'+R;
        K=P*H'/S;
        x=x+K*v;
        P=(I-K*H)*P;
        xOut(:,k)=[ze;x;v;];
    end
    dOut=[xOut;zm;qIn];
```

A.2   FMM(·)

This function contains the prediction model for all DQ filters, including the constant velocity (CV), constant acceleration (CA), and alternate constant acceleration (CA2).

Inputs:

x	state viable
t	step time (sec)
m	prediction model (CV, CA, CA2)

```
function xOut=fMM(xIn,t,m)
%predict the next x based on the previous one
% xIn    input (quaternion data in vector form)
% t      time base in seconds
% m      model to run (CV, CA, CA2)

%model indices
CV=0;
CA=1;
CA2=2;   d=1.5;

%state function
w=xIn(1:3);
a=xIn(4:6);
if (m==CV)
    a=a*0;
end

if (m==CA)
```

```
      w=w+a*t;
end


if (m==CA2)
      w=w+d*a*t;
      a=d*a;
end
xOut=[w;a];
```

A.3   GETHMMR(·)

Construct the Jacobian matrix containing the partial derivatives of measurement function with respect to each state variable.

```
function Hout=GetHmmR(zp,x,t,model,f)
%partial derivative of measurement funciton with respect to state
%   x    estimated state
%   t    step time
%   m    state model (0=CV; 1=CA)
%   f    normal or reduced form
%
CV = 0;
CA = 1;
CA2=2;   d=1.5;

%prevent underflow conditions
for i=1:6
    if (x(i)==0)
        x(i)=1e-50;
    end
end

if (f==0)
    w=x(1:3);
    W=sqrt(w'*w);
    dq=GetDq(w,t);

    Hout=zeros(4,6);
    for i=1:3
        Hout(1,i)=-(t/4)*dq(i+1);
        Hout(2,i)=(w(1)/W^2)*(t*w(i)*dq(1)-dq(i+1))+...
```

```
                (Dirac(i,1)/w(i))*dq(i+1);
            Hout(3,i)=(w(2)/W^2)*(t*w(i)*dq(1)-dq(i+1))+...
                (Dirac(i,2)/w(i))*dq(i+1);
            Hout(4,i)=(w(3)/W^2)*(t*w(i)*dq(1)-dq(i+1))+...
                (Dirac(i,3)/w(i))*dq(i+1);
        end

        if (model == CA)
            Hout(:,4:6)=Hout(:,1:3)*t;
        end

        if (model == CA2)
            Hout(:,4:6)=Hout(:,1:3)*d*t;
        end
    else
        Hout=zeros(3,6);
        delta=1e-6;
        for i=1:3
            xs=x;
            xs(i)=xs(i)+delta;
            xs=fMM(xs,t,CV);
            zs=GetZ(xs(1:3),t,f);
            Hout(:,i)=(zs-zp)/delta;
        end

        if (model == CA)
            for i=1:3
                xs=x;
                xs(i+3)=xs(i+3)+delta;
                xs=fMM(xs,t,CA);
                zs=GetZ(xs(1:3),t,f);
                Hout(:,i+3)=(zs-zp)/delta;
            end
        end
    end
end
```

A.4 GETZ(·)

This function computes the delta quaternion of each frame from the estimated angular velocity.
Inputs:

w	angular velocity (rads/sec^2)
t	time step (sec)

```
function    z=GetZ(w,t,f)
%compute the delta quaternion for a angular velocity (no accelration)
% w    angular velocity data as a string of vectors
% t    time base (seconds)
% f    model (CV or CA)

%models supported
% CV = 0
% CA = 1

w=w(1:3);
for j=1:3
    if (w(j)==0)
        w(j)=1e-50;
    end
end
W=sqrt(w'*w);
theta=W*t/2;
sinTheta=sin(theta);
cosTheta=cos(theta);

if (f==0)
    z=[cosTheta;2*w*sinTheta/W];
    z=z/sqrt(z'*z);
else
    z=(2*w*sinTheta/(W*sqrt(1+3*sinTheta^2)));
end

%force a positive rotation
if (cosTheta<0)
    z=-z;
end
```

APPENDIX B

Multiple Model Delta Quaternion Filter (MMDQ)

B.1 RUNMMDQ2

This function executes a three filter multiple-model Kalman filter based on the DQ architecture. The DQEKF (Appendix A) provides the Kalman filter for this implementation, other supporting functions are found in Appendix D.

```
function dOut=RunMMDQ2(qIn,Q,R,tpm,t)
%a function to run the MMDQ on a data set with revised algorithm
%qIn        measured quaternion data
%Q          process noise tuning
%R          measurement noise tuning
%tpm        transition probablity matrix
%t          time step

%set data location as required
DataDir='c:\example_dir';

N=length(qIn(1,:)');
M=length(tpm(:,1));
CV=0;
CA=1;

if M==3
    model=[CV,CV,CA]';
else
    model=[CV,CA,CA]';
end

f=0;
%prepare the input
dq1=zeros(4,N);
```

```
for k=1:N-1
    dq1(:,k)=Qmul(qIn(:,k+1),qIn(:,k),0,1);
end
dq1(:,N)=dq1(:,N-1);
zm=dq1;

Acv=[[eye(3,3),eye(3,3)*0];[eye(3,3)*0,eye(3,3)*0]];
Aca=[[eye(3,3),eye(3,3)*t];[eye(3,3)*0,eye(3,3)]];

Wcv=[eye(3,3)*t;eye(3,3)*0];
Wca=[eye(3,3)*t*t/2;eye(3,3)*t];

%get A and W
Am=zeros(3,6,6);
WQWm=zeros(3,6,6);
for  i= 1:M
    s1=(i-1)*3+1;
    s2=s1+2;
    if (model(i)==CV)
        Am(i,:,:)=Acv;
        WQWm(i,:,:)=Wcv*Q(s1:s2,:)*Wcv';
    else
        Am(i,:,:)=Aca;
        WQWm(i,:,:)=Wca*Q(s1:s2,:)*Wca';
    end
end

cbar=zeros(3,1);
LLIMIT=10e-50;
sR=sqrt(2*pi)^M;
I=eye(6,6);
x=zeros(6,3);
x0=zeros(6,3);
P=zeros(6,6);
Pm=zeros(3,6,6);
P0=zeros(3,6,6);
Pa=zeros(6,6);
Pb=zeros(6,6);
u=zeros(3,1);
```

```
uP=zeros(3,3);
L=zeros(3,1);
xOut=zeros(34,N);
xe=zeros(6,1);

%initialization
for j=1:M
    x(:,j)=[1e-12,1e-12,1e-12,1e-12,1e-12,1e-12]';
    Pm(j,:,:)=eye(6,6);
    u(j)=1/M;
    L(j)=1/M;
end

%adaptive tpm
tpmK=100;
tpmCnt=0;
gamma=zeros(M,M);
alpha0=zeros(M,M);
alpha=tpm;

aa=1;
aTPM=0;
for k=1:N

if (aTPM==1)
    %gamma
    for i=1:M
        for j=1:M
            gamma(i,j)=...
1+(1/(u(1:M)'*tpm*L(1:M)))*u(i)*(L(j)-tpm(:,i)'*L(1:M));
        end
    end

    %alpha
    for i=1:M
        for j=1:M
            alpha0(i,j)=...
            alpha(i,j)+alpha(i,j)*gamma(i,j)/(alpha(i,:)*gamma(i,:)');
        end
```

```
        end
        alpha=alpha0;

        %tpm switch
        if (tpmCnt==tpmK)
            tpm=(1/tpmK+1)*alpha;

            %make sure it is normalized by model
            for i=1:M
                tpm(i,:)=tpm(i,:)/sum(tpm(i,:));
            end

            %make sure non of the tpm elements are zero
            for i=1:M
                for j=1:M
                    if (tpm(i,j)<LLIMIT)
                        tpm(i,j)=LLIMIT;
                    end
                end
            end

            tpmCnt=0;
            alpha=tpm;
        end
    end
    if (aa==1)
        %mixing probablities
        for j=1:M
            cbar(j)=0;
            for i=1:M
                uP(i,j)=tpm(i,j)*u(i);

                %prevent zero mixing coef
                if (uP(i,j)<LLIMIT)
                    uP(i,j)=LLIMIT;
                end
                cbar(j)=cbar(j)+uP(i,j);
            end
            uP(:,j)=uP(:,j)/cbar(j);
```

```
end

%state Interaction
x0=x0*0;
for j=1:M
   for i=1:M
       x0(:,j)=x0(:,j)+x(:,i)*uP(i,j);
   end
end

for j=1:M
    Pa=Pa*0;
    for i=1:M
        vs=x(:,i)-x0(:,j);
        Pb(:,:)=Pm(i,:,:);
        Pa=Pa+uP(i,j)*(Pb+OuterProduct(vs,vs));
    end
    P0(j,:,:)=Pa;
end
end
%covariance interaction
x=x0;
Pm=P0;

%Kalman Filtering
for j=1:M

    %cnvert to 6x6 to avoid error flags
    P(:,:)=Pm(j,:,:);
    A(:,:)=Am(j,:,:);
    WQW(:,:)=WQWm(j,:,:);

    x(:,j)=fMM(x(:,j),t,model(j));
    ze=GetZ(x(1:3,j),t,f);
    H=GetHmmR(ze,x(:,j),t,model(j),f);
    P=A*P*A'+WQW;
    v=zm(:,k)-ze;
    S=H*P*H'+R;
    K=P*H'/S;
```

```
        x(:,j)=x(:,j)+K*v;
        Pm(j,:,:)=(I-K*H)*P;

        %mode liklihood
        L(j)=(1/(sR*sqrt(det(S))))*exp(-0.5*v'/S*v);
        if (L(j)<LLIMIT)
            L(j)=LLIMIT;
        end
    end

    %mode probability
    c=0;
    for j=1:M
        u(j)=L(j)*cbar(j);

        %prevent zero mixing coef
        if (u(j)<LLIMIT)
            u(j)=LLIMIT;
        end
        c=c+u(j);
    end
    u=u/c;


    %estimate and covariance combination
    xe=xe*0;
    for j=1:M
        xe=xe+x(:,j)*u(j);
    end
%xe=x(:,1);
    dq=GetDq(xe(1:3),t);
    xOut(:,k)=[dq;xe;x(:,1);x(:,2);x(:,3);L;u];

tpmCnt=tpmCnt+1;
end
dOut=[xOut;zm;qIn];

if (M==2)
```

```
    dlmwrite([DataDir,'AdaptivelyFoundTPM2.csv'],tpm);
else
    dlmwrite([DataDir,'AdaptivelyFoundTPM3.csv'],tpm);
end
```

APPENDIX C

Interpolation Volume Calibration (IVC)

The Matlab script (RunStickMapping) runs all functions necessary to generate a map from collected data using the IVC process. Supporting functions for this module follow in this appendix while general purpose functions are found in Appendix D.

C.1   RUNSTICKMAPPING.M

```
%this script runs the stick mapping interpolation

%we reduce the map size to (9, 9, 9) to be able to handle all the points
%without moving sensors around.  This also reduces the amount of data we
%need.
%The map is modified in ProcessCubeData()

%input data is arranged in a root (SimDir) with three subdirectories
%("DataCollection", "InterpolationData", and "LUTdata").  The Data
%collection contains data files created from the output of a Polhemus
%Liberty tracker, "InterpolationData" contains intermediate files and
%"LUTdata" contains the distortion map of the target.

%set paths to your locations
SimDir='SET TO DATA ROOT';
DataColl=cat(2,SimDir,'DataCollection\');
IntData=cat(2,SimDir,'InterpolationData\');
LUTdata=cat(2,SimDir,'LUTdata\');

%this is the cutoff value for the signal matrix error
sval=0.3;

%process the data files into known format
names=caseread(horzcat(DataColl,'DataFileList.txt'));
[Nn,~]=size(names);
```

```
for i=1:Nn
    ProcessDataFile(names(i,:),'data',SimData);
end

%get the cube data
ProcessCubeData_PnO('cube_CubeFixture2.csv',SimDir);

%run the stats so we can create the correction polynomials
[rErr1,aerErr1]=RunInterpStats('ProcessedDataList.txt','Opt',10000,SimDir);

%take off the hish error samples
Opt_stats=dlmread('Opt_stats.csv');
table=SortTableByVal(Opt_stats,4,sval,0);
dlmwrite([IntData,'Opt_stats_decimated.csv'],table);

%get the error polynomial coef
FitCubePoly('Opt_stats_decimated.csv',SimDir);

%process the collected data to create the mapping table
[good,bad]=ProcessPnO_debug('Opt',sval,SimDIr);

%create a table of estimated positions and secondary field
GenerateFieldDataSimDir(SimDir);

%resample to a uniform grid using natural neighbor interpolation
GenerateMapC_NaturalNeighbor(SimDir);

%generate analysis
RunMapStats(SimDir);

RunMapStats_MultiLevel('mild',0.03,SimDir);
RunMapStats_MultiLevel('moderate',0.15,SimDir);
RunMapStats_MultiLevel('severe0',0.45,SimDir);
RunMapStats_MultiLevel('severe1',0.75,SimDir);
RunMapStats_MultiLevel('severe2',1,SimDir);
```

C.2   CREATEMAPSTRUCT(·)

```
function map_struct = CreateMapStruct( lut,min,incr,grid)
%create the mapstruct data type from individual values
%this is done to standardize the format.  The format assumes
%that the map is aligned with source reference frame and the
%map is rectangular in shape using a uniform grid.
% lut    look up table of measured field values
% min    minimum position of table
% incr   increment between grid points
% grid   grid size of table

map_struct.lut=lut;
map_struct.Min=min;
map_struct.Incr=incr;
map_struct.Grid=grid;

%create the max values
for i=1:3
    map_struct.Max(i) = map_struct.Min(i) +...
        (map_struct.Grid(i)-1)*map_struct.Incr(i);
end

end
```

C.3   FITCUBEPOLY(·)

```
function FitCubePoly(sFile,SimDir)
%FitCubePoly()  generate the error correction and error estimation
%               polynomials for the cube Pn0 estimates
%sFile          cube statics filename
%SimDir         root directory of data directories

%set directories
DataColl=cat(2,SimDir,'DataCollection\');
IntData=cat(2,SimDir,'InterpolationData\');
LUTdata=cat(2,SimDir,'LUTdata\');
stats=dlmread([IntData,sFile]);
[N,~]=size(stats);
```

```
%find the end of the file by looking for 0 in the index (first) column
m=0;
for n=1:N
    if (stats(n,1)>0)
        m=m+1;
    end
end
N=m;
stats=stats(1:N,:);

%break out data for S1
s4vec_err=stats(:,5:13)';
rTst=stats(:,14:16)';
qTst=stats(:,17:20)';
rEst=stats(:,21:23)';
qEst=stats(:,24:27)';
rErrT=stats(:,2);
aerErrT=stats(:,3);
serr=stats(:,4);

%Generate errors
for n=1:N
    qEst(:,n)=qEst(:,n)/sqrt(qEst(:,n)'*qEst(:,n));
end

%generate the polyterms for each point
svTerms=zeros(91,N);%(455,N);%(91,N);
for n=1:N
    terms=[rEst(:,n);s4vec_err(:,n)];
    a=ComputePoly(terms,2);
    svTerms(:,n)=a;
end

%generate position correction polynomial
rCoef=mvregress(svTerms',rErrT);
aerCoef=mvregress(svTerms',aerErrT);

%check the errors
rErrTest=svTerms'*rCoef;
```

```
aerErrTest=svTerms'*aerCoef;

dlmwrite(horzcat(IntData,'rCoef.csv'),rCoef);
dlmwrite(horzcat(IntData,'aerCoef.csv'),aerCoef);

ErrOut=[rTst',rEst',rErrTest,rErrT,qTst',qEst',aerErrTest,aerErrT];
dlmwrite([IntData,'PolyFit_ErrEst.csv'],ErrOut);
dlmwrite([IntData,'svTerms4.csv'],svTerms');
end
```

C.4   GENERATEFIELDDATA(·)

```
function    GenerateFieldData(SimDir)
%Generate a uniform grid field map from a data cloud using Matlab nearest
%neighbor interpolation

%set paths
DataColl=cat(2,SimDir,'DataCollection\');
IntData=cat(2,SimDir,'InterpolationData\');
LUTdata=cat(2,SimDir,'LUTdata\');

%these are the original map adjusted to the interpolation volume
% LUT.csv is the look up table of measured distortin in the map area
% param.csv contains the minimum, increment and grid values for the LUT
lutM=dlmread(horzcat(LUTdata,'LUT.csv'));
paramM=dlmread([LUTdata,'param.csv']);
minM=paramM(:,1);
incrM=paramM(:,2);
gridM=paramM(:,3);
MapStructM=CreateMapStruct(lutM,minM,incrM,gridM);

%read the pno data file
% 'DataCloud.csv' contains the measured data.
pnoData=dlmread(horzcat(DataColl,'DataCloud.csv'));
[N,M]=size(pnoData);
OutData=zeros(N,12);

%Generate the signal matrix for each measured point
for i=1:N
```

```
    %get the estimated PnO of the next measurement
    rEst=pnoData(i,1:3);
    qEst=pnoData(i,4:7);
    Aest=QtoA(qEst);

    rMeas=pnoData(i,8:10);
    qMeas=pnoData(i,11:14);
    Ameas=QtoA(qMeas);

    %get the field measurement
        %get the unrotated measured field
        svm=GetField(rMeas',MapStructM);

        %rotate measured field ito measured aer
        smeas=Ameas'*[svm(1:3)';svm(4:6)';svm(7:9)'];

    %compute the distortion based on estimated PnO
        %get ideal field at estimated positon
        sideal=GetIdealField(rEst');

        %get the unrotated distortion matrix
        gmeas=Aest*smeas-sideal;

        %convert to a vector
        sval=[gmeas(1,:) gmeas(2,:) gmeas(3,:)];

    OutData(i,:)=[rEst sval];
end

%write data to file

dlmwrite([DataColl,'SampledFieldData.csv'],OutData);
```

C.5   GENERATEMAPC_NATURALNEIGHBOR(·)

```
function    GenerateMapC_NaturalNeighbor(SimDir)
%Generate a uniform grid field map from a data cloud using Matlab nearest
%neighbor interpolation
%
%SimDir       root directory of data storage
```

```
%set path to data file locations
DataColl=cat(2,SimDir,'DataCollection\');
IntData=cat(2,SimDir,'InterpolationData\');
LUTdata=cat(2,SimDir,'LUTdata\');

%Make sure the maps align correctly with the interpolation cube ideal
%positions and build the map locations off the interpolation volume
%positions
%IdealCuve_xyz.csv contains the ideal cube indices.
Idealxyz=dlmread([IntData,'IdealCube_xyz.csv']);
CubeCenter(1)=Idealxyz(1)+0.5*(Idealxyz(2)-Idealxyz(1));
CubeCenter(2)=Idealxyz(3)+0.5*(Idealxyz(4)-Idealxyz(3));
CubeCenter(3)=Idealxyz(5)+0.5*(Idealxyz(6)-Idealxyz(5));

%these are the original map adjusted to the interpolation volume
% get the mapped distortion
lutM=dlmread(horzcat(LUTdata,'LUT.csv'));

%get the map paramters (min, incr, grid)

%create the MapStruct
paramM=dlmread([LUTdata,'param.csv']);
minM=paramM(:,1);
incrM=paramM(:,2);
gridM=paramM(:,3);
MapStructM=CreateMapStruct(lutM,minM,incrM,gridM);

%this is the target map
minS(1,1)=CubeCenter(1)-2.5*incrM(1);
minS(2,1)=CubeCenter(2)-2.5*incrM(1);
minS(3,1)=CubeCenter(3)-2.5*incrM(1);
incrS=incrM;
gridS=[6,6,6]';
Ns=gridS(1)*gridS(2)*gridS(3);
paramS=[minS incrS gridS];

%get the field data
fData=dlmread(horzcat(DataColl,'SampledFieldData.csv'));
```

```
xyz=fData(:,1:3);
sval=fData(:,4:12);

%create the DelaunayTri object for the interpolation
%DT=DelaunayTri(xyz);
F1=TriScatteredInterp(xyz,sval(:,1),'natural');
F2=TriScatteredInterp(xyz,sval(:,2),'natural');
F3=TriScatteredInterp(xyz,sval(:,3),'natural');
F4=TriScatteredInterp(xyz,sval(:,4),'natural');
F5=TriScatteredInterp(xyz,sval(:,5),'natural');
F6=TriScatteredInterp(xyz,sval(:,6),'natural');
F7=TriScatteredInterp(xyz,sval(:,7),'natural');
F8=TriScatteredInterp(xyz,sval(:,8),'natural');
F9=TriScatteredInterp(xyz,sval(:,9),'natural');

%create the grid points
xi=zeros(gridS(1));
yi=zeros(gridS(2));
zi=zeros(gridS(3));
for i=1:gridS(1)
    xi(i)=minS(1)+(i-1)*incrS(1);
end
for i=1:gridS(2)
    yi(i)=minS(2)+(i-1)*incrS(2);
end
for i=1:gridS(3)
    zi(i)=minS(3)+(i-1)*incrS(3);
end
[x,y,z]=meshgrid(xi,yi,zi);

%create the new field data
sval1=F1(x,y,z);
sval2=F2(x,y,z);
sval3=F3(x,y,z);
sval4=F4(x,y,z);
sval5=F5(x,y,z);
sval6=F6(x,y,z);
sval7=F7(x,y,z);
sval8=F8(x,y,z);
```

```
sval9=F9(x,y,z);

cnt=1;
lutS=zeros(Ns,9);
for i=1:gridS(1)
    for j=1:gridS(2)
        for k=1:gridS(3)
            lutS(cnt,:)=[sval1(i,j,k) sval2(i,j,k) sval3(i,j,k)...
                sval4(i,j,k) sval5(i,j,k) sval6(i,j,k) sval7(i,j,k)...
                sval8(i,j,k) sval9(i,j,k)];
            cnt=cnt+1;
        end
    end
end

%test it
sGridS=zeros(Ns,9);
for n=1:Ns
    sGridS(n,:)=Lookup(xyz(n,:)',MapStructM)';
end

%write data to file
dlmwrite([DataColl,'SampledLut.csv'],lutS);
dlmwrite([DataColl,'SampledLutParam.csv'],paramS);
```

C.6   PROCESSCUBEDATA_PNO(·)

```
function ProcessCubeData_Pn0(InFile,SimDir)
%ProcessCubeData(): generate cube positons and field measurements
%fData:      filename containing UTH mode data
%fChar:      filename containing characterization without extraneous
%            characters
%Nindex:     sensor reordering to assemble XYZ organization
%SimDir      root location of data directories

%set paths
DataColl=cat(2,SimDir,'DataCollection\');
IntData=cat(2,SimDir,'InterpolationData\');
LUTdata=cat(2,SimDir,'LUTdata\');
```

```
MET2IN=39.3701;
IN2MET=1/MET2IN;
RAD2DEG=180/pi;
DEG2RAD=1/RAD2DEG;

%bring in the data and process it
table=dlmread(InFile);
[N,~]=size(table);
%The input data has been preprocessed to place the data in rows of eight
%sensors

%get the lut so we can add distortion to the undistorted measured data
%LUT.csv is the distortion map
%param are the map parameters (min,incr,grid)
lut=dlmread([LUTdata,'LUT.csv']);
param=dlmread([LUTdata,'param.csv']);
min=param(:,1);
incr=param(:,2);
grid=param(:,3);

sensors=8;

rMeas=zeros(sensors,3,N);
aerMeas=zeros(sensors,3,N);
rA=zeros(3,8);
aerA=zeros(3,8);
rS=zeros(3,8);
aerS=zeros(3,8);
for i=1:sensors
    for n=1:N
        %get PnO
        rStart=(i-1)*3+1;
        qStart=(i-1)*4+25;
        rMeas(i,:,n)=table(n,rStart:rStart+2)';
        aerMeas(i,:,n)=QtoAer(table(n,qStart:qStart+3)');
    end

    for j=1:3
        rA(j,i)=mean(rMeas(i,j,:));
```

```
        rS(j,i)=std(rMeas(i,j,:));

        aerA(j,i)=mean(aerMeas(i,j,:));
        aerS(j,i)=std(aerMeas(i,j,:));
    end
end

%switch the sensors into XYZ order and finalize average
Nindex=[3 4 1 2 7 8 5 6];

%compute the relative positions to determine what the ordering is
rAvg=zeros(3,8);
aerAvg=zeros(3,8);
for i=1:sensors
    rAvg(:,Nindex(i))=rA(:,i);
    aerAvg(:,Nindex(i))=aerA(:,i);
end

%validate teh ordering
rP=zeros(3,sensors);
for i=1:sensors
    rP(:,i)=rAvg(:,i)-rAvg(:,1);
end

%create the cube fixture frame and a transformation matrix to move from
%global to local.  This will allow up to detemine the relative positions
%of the measured sensor positions in any rotation.
ra=rAvg(:,5)-rAvg(:,1);
rb=rAvg(:,3)-rAvg(:,1);
rx=ra/sqrt(ra'*ra);
rz=cross(ra,rb);
rz=rz/sqrt(rz'*rz);
ry=cross(rz,rx);
ry=ry/sqrt(ry'*ry);
XYZ=[rx,ry,rz];
xyz=[1 0 0;0 1 0; 0 0 1];
Tab=zeros(3,3);
Tab(:,1)=[dot(xyz(:,1),XYZ(:,1)),dot(xyz(:,2),XYZ(:,1)),...
            dot(xyz(:,3),XYZ(:,1))]';
```

```
Tab(:,2)=[dot(xyz(:,1),XYZ(:,2)),dot(xyz(:,2),XYZ(:,2)),...
               dot(xyz(:,3),XYZ(:,2))]';
Tab(:,3)=[dot(xyz(:,1),XYZ(:,3)),dot(xyz(:,2),XYZ(:,3)),...
               dot(xyz(:,3),XYZ(:,3))]';
Tba=Tab';

%Now rotate the cube into alignment with the source frame
rAvg2=zeros(3,sensors);
aerAvg2=zeros(3,sensors);
for i=1:8
    rAvg2(:,i)=Tba*rAvg(:,i);
    aerAvg2(:,i)=AtoAer(Tba*AerToA(aerAvg(:,i)));
end

%save so we can optimize in Mathcad for quick turn around
%dlmwrite([IntData,'AverageCube.csv'],rAvg2);

%read back optimized solution from Mathcad
%this method is justified by the premise of our simulation, the map and
%the cube must align to give repeatable results.
rPopt=dlmread([IntData,'OptimizedCube.csv']);
rPfly=dlmread([IntData,'FlyCube.csv']);
rPideal=dlmread([IntData,'IdealCube.csv']);

%Determine the Map offset to align it with the interpolation cube.
rC=zeros(3,1);
for i=1:8
    rC=rC+rAvg2(:,i);
end
rC=rC/sensors;
%we will put the interpolation volume in cell (7,6,4) of the LUT map
Min2=rC-([7*incr(1);6*incr(2);4*incr(3)]+0.5*incr);
dlmwrite([LUTdata,'LUTparam.csv'],[Min2,incr,grid]);

%create a new map profile for our output of a (6,6,6) map centered at rC
Min3=rC-[2.5*incr(1);2.5*incr(2);2.5*incr(3)];
Incr3=incr;
Grid3=[6 6 6]';
param3=[Min3,Incr3,Grid3];
```

```
dlmwrite([LUTdata,'param3.csv'],param3);

%Get the S4 standard deviation
s4noise_std=dlmread([IntData,'CubeInterpFixture_S4_stdev.csv']);
s4noise_mean=[0 0 0 0 0 0 0 0 0]';

%Now compute the relative positions
delta=[0.0381;0.0381;0.0381];
rOpt=zeros(3,sensors);
rFly=zeros(3,sensors);
rIdeal=zeros(3,8);
aerOpt=zeros(3,sensors);
qOpt=zeros(4,sensors);
qPopt=zeros(4,sensors);
OptVals=zeros(9,sensors);
FlyVals=zeros(9,sensors);
IdealVals=zeros(9,sensors);
for i=1:sensors
    rOpt(:,i)=rPopt(:,i)+rC-0.5*delta;
    rFly(:,i)=rPfly(:,i)+rC-0.5*delta;
    rIdeal(:,i)=rPideal(:,i)+rC-0.5*delta;

    q=Qmul(AerToQ(aerAvg2(:,i)),AerToQ(aerAvg2(:,1)),0,1);
    qPopt(:,i)=q/sqrt(q'*q);
    q=AerToQ(AerToQ(aerAvg2(:,i)));
    qOpt(:,i)=q/sqrt(q'*q);
    aerOpt(:,i)=QtoAer(qOpt(:,i));

    OptVals(:,i)=GetField(rOpt(:,i),lut,Min2,incr,grid)+...
            random('norm',s4noise_mean,s4noise_std);
    FlyVals(:,i)=GetField(rFly(:,i),lut,Min2,incr,grid)+...
            random('norm',s4noise_mean,s4noise_std);
    IdealVals(:,i)=GetField(rIdeal(:,i),lut,Min2,incr,grid)+...
            random('norm',s4noise_mean,s4noise_std);
end
qFly=qOpt;
qPfly=qPopt;
aerFly=aerOpt;
```

```
qIdeal=qOpt;
qPideal=qPopt;
aerIdeal=aerOpt;

OptXyz=[rOpt(1,1);rOpt(1,5);rOpt(2,1);rOpt(2,3);rOpt(3,1);rOpt(3,2)];
FlyXyz=[rFly(1,1);rFly(1,5);rFly(2,1);rFly(2,3);rFly(3,1);rFly(3,2)];
IdealXyz=[rIdeal(1,1);rIdeal(1,5);rIdeal(2,1);rIdeal(2,3);...
            rIdeal(3,1);rIdeal(3,2)];
xyz=[rIdeal(1,1);rIdeal(1,5);rIdeal(2,1);rIdeal(2,3);rIdeal(3,1);...
            rIdeal(3,2)];

%test the ideal case
rIdealEst=zeros(3,sensors);
rIdealErr=zeros(3,8);
aerIdealEst=zeros(3,sensors);
aerIdealErr=zeros(3,8);
qSeed=[1 0 0 0]';
for i=1:8
    svec=GetField(rIdeal(:,i),lut,Min2,incr,grid);
    s4=AerToA(QtoAer(qIdeal(:,i)))'*[svec(1:3)';svec(4:6)';svec(7:9)'];
    s4vec=[s4(1,:),s4(2,:),s4(3,:)]';

    %find the solution
    rIdealEst(:,i)=rFindR(rC,s4vec,xyz,IdealVals');
    svest=Interp(rIdealEst(:,i),xyz,IdealVals');
    sest=[svest(1:3);svest(4:6);svest(7:9)];
    qIdealEst=SolveForOrientation(sest,s4);
    aerIdealEst(:,i)=QtoAer(qIdealEst);

    %compute errors
    rIdealErr(:,i)=rIdealEst(:,i)-rIdeal(:,i);
    aerIdealErr(:,i)=aerIdealEst(:,i)-aerIdeal(:,i);
end

%test the optimized cube
rOptEst=zeros(3,sensors);
rOptErr=zeros(3,8);
aerOptEst=zeros(3,sensors);
aerOptErr=zeros(3,8);
```

```
for i=1:8
    svec=GetField(rOpt(:,i),lut,Min2,incr,grid);
    s4=AerToA(QtoAer(qOpt(:,i)))'*[svec(1:3)';svec(4:6)';svec(7:9)'];
    s4vec=[s4(1,:),s4(2,:),s4(3,:)]';

    %find the solution
    rOptEst(:,i)=rFindR(rC,s4vec,xyz,OptVals');
    svest=Interp(rOptEst(:,i),xyz,OptVals');
    sest=[svest(1:3);svest(4:6);svest(7:9)];
    qOptEst=SolveForOrientation(qSeed,sest,s4);
    aerOptEst(:,i)=QtoAer(qOptEst);

    %compute errors
    rOptErr(:,i)=rOptEst(:,i)-rOpt(:,i);
    aerOptErr(:,i)=aerOptEst(:,i)-aerOpt(:,i);
end

%test the Fly cube
rFlyEst=zeros(3,sensors);
rFlyErr=zeros(3,8);
aerFlyEst=zeros(3,sensors);
aerFlyErr=zeros(3,8);
for i=1:8
    svec=GetField(rFly(:,i),lut,Min2,incr,grid);
    s4=AerToA(QtoAer(qFly(:,i)))'*[svec(1:3)';svec(4:6)';svec(7:9)'];
    s4vec=[s4(1,:),s4(2,:),s4(3,:)]';

    %find the solution
    rFlyEst(:,i)=rFindR(rC,s4vec,xyz,FlyVals');
    svest=Interp(rFlyEst(:,i),xyz,FlyVals');
    sest=[svest(1:3);svest(4:6);svest(7:9)];
    qFlyEst=SolveForOrientation(qSeed,sest,s4);
    aerFlyEst(:,i)=QtoAer(qFlyEst);

    %compute errors
    rFlyErr(:,i)=rFlyEst(:,i)-rFly(:,i);
    aerFlyErr(:,i)=aerFlyEst(:,i)-aerFly(:,i);
end
```

```
svec=GetField(rC,lut,Min2,incr,grid);
aerc=QtoAer(qFly(:,4));
s4=AerToA(aerc)'*[svec(1:3)';svec(4:6)';svec(7:9)'];
s4vec=[s4(1,:),s4(2,:),s4(3,:)]';
rcVals=[IdealVals,OptVals,FlyVals];
rcEst=zeros(3,3);
rcErr=zeros(3,3);
qcEst=zeros(4,3);
aercErr=zeros(3,3);
for i=1:3
    rcEst(:,i)=rFindR(rC,s4vec,xyz,rcVals(:,(i-1)*8+1:(i-1)*8+8)');
    svest=Interp(rcEst,xyz,rcVals(:,(i-1)*8+1:(i-1)*8+8)');
    sest=[svest(1:3);svest(4:6);svest(7:9)];
    qcEst(:,i)=SolveForOrientation(qSeed,sest,s4);

    %compute errors
    rcErr(:,i)=rcEst(:,i)-rC;
    aercErr(:,i)=QtoAer(Qmul(qFly(:,4),qcEst(:,i),0,1));
end

%save the ideal results
dlmwrite(horzcat(IntData,'IdealCube_PnO.csv'),vertcat(rIdeal,qIdeal));
dlmwrite(horzcat(IntData,'IdealCube_rPnO.csv'),vertcat(rPideal,qPideal));
dlmwrite(horzcat(IntData,'IdealCube_xyz.csv'),IdealXyz);
dlmwrite(horzcat(IntData,'IdealCube_vals.csv'),IdealVals);

%save the Opt results
dlmwrite(horzcat(IntData,'OptCube_PnO.csv'),vertcat(rOpt,qOpt));
dlmwrite(horzcat(IntData,'OptCube_rPnO.csv'),vertcat(rPopt,qPopt));
dlmwrite(horzcat(IntData,'OptCube_xyz.csv'),OptXyz);
dlmwrite(horzcat(IntData,'OptCube_vals.csv'),OptVals);

%save the Fly results
dlmwrite(horzcat(IntData,'FlyCube_PnO.csv'),vertcat(rFly,qFly));
dlmwrite(horzcat(IntData,'FlyCube_rPnO.csv'),vertcat(rPfly,qPfly));
dlmwrite(horzcat(IntData,'FlyCube_xyz.csv'),FlyXyz);
dlmwrite(horzcat(IntData,'FlyCube_vals.csv'),FlyVals);

end
```

C.7 PROCESSDATAFILE(·)

This function reads data files, formatting them for use with the stick mapping environment.

```
function [d,cnt]=ProcessDataFile(filename,prefix,SimDir)
%   CreateIntStatData()     organize the tracker output data into one
%                           column per frame
%   filename    data file name
%   prefix      prefix to apply to output files
%   SimDir      root location of data directories

%set paths
DataColl=cat(2,SimDir,'DataCollection\');
IntData=cat(2,SimDir,'InterpolationData\');
LUTdata=cat(2,SimDir,'LUTdata\');

len=length(filename);
for i=1:10
    if (filename(len)=='.')
        break;
    else
        len=len-1;
    end
end
pOutFile=[prefix,'_pno_',filename(1:len),'.csv'];
sOutFile=[prefix,'_smeas_',filename(1:len),'.csv'];

MET2IN=39.3701;
IN2MET=1/MET2IN;
RAD2DEG=180/pi;
DEG2RAD=1/RAD2DEG;

if strncmp(prefix,'cube',4)
    sensors=8;
else
    sensors=9;
end
fid=fopen([DataColl,filename]);

[dTable,Nd]=fscanf(fid,'%u %f %f %f %f %f %f');
Nd=floor(Nd/7);
```

```
N=floor(Nd/sensors);
dOut=zeros(N,sensors*7);
rM=zeros(1,sensors*3);
qM=zeros(1,sensors*4);
table=zeros(sensors,7);

%if strncmp(prefix,'data',4)
%    N=100;
%end
n=1;

cnt=0;
d=1;
while (n<=N)
    %the default case is data
    if strncmp(filename,'alt1_',4)
        %sensor 0 is in position two so swap them
        for j=1:sensors
            start=(n-1)*sensors*7+(j-1)*7+1;
            table(((sensors+1)-j),:)=dTable(start:start+6,1)';
        end
    else
        %standard data types
        for j=1:sensors
            start=(n-1)*sensors*7+(j-1)*7+1;
            table(j,:)=dTable(start:start+6,1)';
        end
    end

    %get Measured PnO
    for j=1:sensors
        rStart=(j-1)*3+1;
        qStart=(j-1)*4+1;
        rM(rStart:rStart+2)=table(j,2:4)*IN2MET;
        q=AerToQ(table(j,5:7)'*DEG2RAD);
        q=q/sqrt(q'*q);
        qM(qStart:qStart+3)=q;

    end
```

```
    dOut(d,:)=[rM,qM];
    d=d+1;

    n=n+1;

    %generate the signal matrix for sensor 0
    s=GetField(rM(1:3)',lut,Min,Incr,grid);
    aer=QtoAer(qM(1:4)');
    A=AerToA(aer);
    sm=A'*[s(1:3)';s(4:6)';s(7:9)'];
    sMvec(d,:)=[sm(1,:),sm(2,:),sm(2,:)];


end
d=d-1;
dlmwrite([DataColl,pOutFile],dOut(1:d,:));
dlmwrite([DataColl,sOutFile],sMvec(1:d,:));

end
```

C.8   PROCESSPNO(·)

```
function    [good,bad]=ProcessPn0(prefix,sval,SimDir)
%ProcessPn0()   process position data to create data table for mapping
%
% prefix      prefix to apply to output files
% sval        an error limit for included data
% SimDir      root location of data directories

%set paths
DataColl=cat(2,SimDir,'DataCollection\');
IntData=cat(2,SimDir,'InterpolationData\');
LUTdata=cat(2,SimDir,'LUTdata\');

%declare local variables
step=10;                   %set the data decimation ration
sensors=9;                 %set the number of sensors on the stick

%declare variable
```

```
rEst=zeros(3,sensors);
qEst=zeros(4,sensors);

rMeas=zeros(3,sensors);
qMeas=zeros(4,sensors);
aerMeas=zeros(3,sensors);

SenErr=zeros(sensors,1);
%OutBuff=zeros(outsize*sensors,15);
%errBuff=zeros(outsize,17);
dOut=zeros(16e+06,15);
errOut=zeros(2e+06,17);

rErrS=zeros(3,sensors);
aerErrS=zeros(3,sensors);
qErrS=zeros(4,sensors);
aerErrST=zeros(sensors,1);
rErrST=zeros(sensors,1);
rE=zeros(3,sensors);
qE=zeros(4,sensors);
rEstM=zeros(3,sensors);
qEstM=zeros(4,sensors);

xyz=dlmread([IntData,'IdealCube_xyz.csv']);
rc=[xyz(1)+0.5*(xyz(2)-xyz(1));xyz(3)+0.5*(xyz(4)-xyz(3));
        xyz(5)+0.5*(xyz(6)-xyz(5))];

%get the fixture noise standard deviation
sdev=dlmread(horzcat(IntData,'CubeInterpFixture_S4_stdev.csv'));

%bring in the lut and lut parameters
lutM=dlmread(horzcat(MatLabSim,'LUTdata\LCDmap_lut.csv'));
paramM=dlmread([LUTdata,'LCDmap_param.csv']);
minM=paramM(:,1);
incrM=paramM(:,2);
gridM=paramM(:,3);
MapStructM=CreateMapStruct(lutM,minM,incrM,gridM);

%the vals are cube dependent so these change with the prefix
```

```
rOpt=dlmread([IntData,prefix,'Cube_PnO.csv']);
rOpt=rOpt(1:3,:);
for i=1:8
    vals(:,i)=GetField(rOpt(:,i),MapStructM);
end

%get correction coef
rCoef=dlmread(horzcat(IntData,'rCoef.csv'));
aerCoef=dlmread(horzcat(IntData,'aerCoef.csv'));

%The PnO file names MUST appear in the list below and have the same number
%of columns (add spaces to equalize the rows, the spaces will be removed
%before the file is read).
names=[
'data_std_TV.csv    ';
'data_std_TH.csv    ';
'data_alt4_516.csv ';
'data_alt1_BH.csv  ';
'data_alt1_BV.csv  ';
'data_alt2_TT.csv  ';
'data_alt3_BV.csv  ';
'data_alt3_CC.csv  ';
'data_alt3_TV.csv  ';
'data_alt3_TH.csv  ';
'data_alt3_TT.csv  ';
'data_alt5_post.csv'
];
[Nn,Mn]=size(names);

%init the seeds
rSeed=[xyz(1,1)+0.5*(xyz(2,1)-xyz(1,1));xyz(3,1)+0.5*(xyz(4,1)-xyz(3,1));
       xyz(5,1)+0.5*(xyz(6,1)-xyz(5,1))];
%qSeed(:,1)=[1 0 0 0]';
good=0;
BadS=0;
BadFlag=0;
count=1;                    %init sample counter
p=1;
ecnt=1;
```

```
cMin=[xyz(1);xyz(3);xyz(5)];
cMax=[xyz(2);xyz(4);xyz(6)];

for i=1:Nn
    %generate data filename with path from list;
    j=0;
    for ai=1:Mn
        if ((names(i,ai)==' ')&&(j<=Mn))
            break;
        else
            j=j+1;
        end
    end
    fName=names(i,1:j);
    filename=[DataColl fName];
    table=dlmread(filename);
    [Nd,~]=size(table);
    N=floor(Nd/step)*step;
    m=1;                        %output record row index

    %shift the median of the position into the center of the cube
    %so that we get SOME good points out of each data set.  We need to do
    %this because we gathered the data on two different days.  All the
    %data collected at the same time does not have this problem.  We use
    %the median to limit the effect of a few bad values.
    if (N>= 1000)
        K=1000;
    else
        K=N;
    end
    rc1=median(table(1:K,1:3));
    delta=rc1'-rc;
    %the data is placed in a common format generate the realtive
    %placements on the stick

    %need to call a special fixture file for each one.
    [rP,qP]=GetRelativeStickSen_Pn0(table(1:K,:),fName);
```

```
%get the magnitude of the sensor offset for error extrapolation
sR=zeros(sensors,1);
for j=1:sensors
    sR(j,1)=sqrt(rP(1:3,j)'*rP(1:3,j));
end


n=1;
while (n<N)
%while(n<10)
    %get the sensor 1 measured Pn0
        for j=1:sensors
            rStart=(j-1)*3+1;
            qStart=(j-1)*4+28;
            rMeas(:,j)=table(n,rStart:rStart+2)'-delta;
            qMeas(:,j)=table(n,qStart:qStart+3)';
            aerMeas(:,j)=QtoAer(qMeas(:,j));
        end

        %process data point if it is in the interpolation volume to save
        %time
        flag=0;
        for j=1:3
            if (rMeas(j,1)<cMin(j))
                flag=1;
            end
            if (rMeas(j,1)>cMax(j))
                flag=1;
            end
        end

    if (flag==0)
        %we pull the measured field from the LUT to simulate how measured
        %values will be different from those interpolated from the cube
        %vertices.  The LUT values are averaged to reduce noise, so we add
        %noise derived from the cube measurements to simulated true data
        svec=GetField(rMeas(:,1),MapStructM);
        snoise=0*sdev.*randn(9,1);
        svec=svec+snoise;
        smeas=[svec(1:3)';svec(4:6)';svec(7:9)'];
```

```
s4meas=AerToA(0*aerMeas(:,1))'*smeas;
s4vec_meas=[s4meas(1,:),s4meas(2,:),s4meas(3,:)]';

%solve for position of sensor 1 in interpolation volume
rEst(:,1)=rFindR(rSeed,s4vec_meas,xyz,vals');
svec_est=Interp(rEst(:,1),xyz,vals')';
sest=[svec_est(1:3)';svec_est(4:6)';svec_est(7:9)'];
qEst(:,1)=qFindQ(sest,s4meas);
qEst(:,1)=qEst(:,1)/sqrt(qEst(:,1)'*qEst(:,1));
aerEst1=QtoAer(qEst(:,1));
Aest=AerToA(aerEst1);

%generate S4vec difference
s4_err=sest-Aest*s4meas;
s4vec_err=[s4_err(1,:),s4_err(2,:),s4_err(3,:)]';
serr=sqrt(s4vec_err'*s4vec_err);

if (serr<sval)
    good=good+1;
  %correct the PnO estimate
    svTerm=ComputePoly([rEst(:,1);s4vec_err],2);

    %log the errors
    rErr=rEst(:,1)-rMeas(:,1);
    aerErr=QtoAer(Qmul(qEst(:,1),qMeas(:,1),0,1));
    rErrT=sqrt(rErr'*rErr);
    aerErrT=sqrt(aerErr*aerErr');

    %positon error prediction
    rErrTp=svTerm'*rCoef;

    %orientation error prediction
    aerErrTp=svTerm'*aerCoef;

    %extend the error to the other sensors
    for j=1:sensors
        SenErr(j)=rErrTp+ sR(j,1)*abs(tan(aerErrTp));
    end
```

```
%back the fixture frame out of the sensor 1 quaternion
qFix=Qmul(qEst(:,1),qP(:,1),0,1);
Afix=AerToA(QtoAer(qFix));
for j=1:sensors
    %position of
    rEst(:,j)=rEst(:,1)+Afix*rP(:,j);

    %orientation of sensor in global frame
    qEst(:,j)=Qmul(qFix,qP(:,j),0,0);
    qEst(:,j)=qEst(:,j)/sqrt(qEst(:,j)'*qEst(:,j));
end

%project the measured PnO from the sensor 0 measurements
qFixM=Qmul(qMeas(:,1),qP(:,1),0,1);
AfixM=AerToA(QtoAer(qFixM));
for j=1:sensors
    %position of
    rEstM(:,j)=rMeas(:,1)+AfixM*rP(:,j);

    %orientation of sensor in global frame
    qEstM(:,j)=Qmul(qFixM,qP(:,j),0,0);
    qEstM(:,j)=qEstM(:,j)/sqrt(qEstM(:,j)'*qEstM(:,j));
end

%check the errors
for j=1:sensors
    rErrS(:,j)=rEst(:,j)-rMeas(:,j);
    rErrST(j,1)=sqrt(rErrS(:,j)'*rErrS(:,j));

    qErrS(:,j)=Qmul(qEst(:,j),qMeas(:,j),0,1);
    aerErrS(:,j)=QtoAer(qErrS(:,j));
    aerErrST(j,1)=sqrt(aerErrS(:,j)'*aerErrS(:,j));

    rE(:,j)=rMeas(:,1)+Afix*rP(:,j);
    qE(:,j)=Qmul(qFix,qP(:,j),0,0);
    qE(:,j)=qE(:,j)/sqrt(qE(:,j)'*qE(:,j));
end

%create an output block for these 8 sensors and append it to
```

```
                    %the global output data
                    dOut(p:p+sensors-1,:)=[rEst',qEst',SenErr,rEstM',qEstM'];

                    %save errors for analysis
                    errOut(ecnt,:)=[rMeas(:,1)',qMeas(:,1)',rErr,aerErr,rErrT,
                            rErrTp,aerErrT,aerErrTp];

                    %recursive seeing
                    rSeed=rEst(:,1);

                    p=p+sensors;
                    ecnt=ecnt+1;          %err buff counter
                    m=m+sensors;          %data buff counter
                    count=count+1;        %increment processing pointer
                else
                    BadS=BadS+1;
                end
            else
                BadFlag=BadFlag+1;
            end
                n=n+step;                 %increment data file pointer
            end
    end
        %write data to file
        dlmwrite([DataColl,'StickMap_DataCloudA.csv'],dOut(1:p-1,:));
        dlmwrite([DataColl,'StickMap_errsA.csv'],errOut(1:ecnt-1,:));

        bad=BadS+BadFlag;

end
```

C.9 RUNINTERPSTATS(·)

```
function [rM,aerM]=RunInterpStats(filename,prefix,N,SimDir)
%RunInterpStats():  Generate errors for interpolation of N random points
%                   in the interpolation cube defined by xyz with vals for
%                   signal matrices (vectors).
%filename       input file name
%prefix         prefix to apply to output files
%N              number of points
```

%SimDir   root location of data directories

```
%set paths
DataColl=cat(2,SimDir,'DataCollection\');
IntData=cat(2,SimDir,'InterpolationData\');
LUTdata=cat(2,SimDir,'LUTdata\');


%get the fixture noise standard deviation
s_stdev=dlmread([IntData,'CubeInterpFixture_S4_stdev.csv']);
snoise_mean=[0 0 0 0 0 0 0 0]';

%we always use the ideal XYZ for interpolation
xyz=dlmread([IntData,'IdealCube_xyz.csv']);
rc=[xyz(1)+0.5*(xyz(2)-xyz(1));xyz(3)+0.5*(xyz(4)-xyz(3));
             xyz(5)+0.5*(xyz(6)-xyz(5))];
%bring in the lut
%here we use param 2 which has been modified to align the LUT with the
%interpolation fixture  param3 is for the new LUT we are creating
lut=dlmread([LUTdata,'LCDmap_lut.csv']);
%these are the original map adjusted to the interpolation volume
param=dlmread([LUTdata,'param2.csv']);
incr1=param(:,2);
grid1=param(:,3);
min1(1,1)=xyz(1)-6*incr1(1);
min1(2,1)=xyz(3)-6*incr1(1);
min1(3,1)=xyz(5)-4*incr1(1);


%the vals are cube dependent so these change with the prefiz
rCube=dlmread([IntData,prefix,'Cube_Pn0.csv']);
rCube=rCube(1:3,:);
vals=zeros(9,8);
for i=1:8
    vals(:,i)=GetField(rCube(1:3,i),lut,min1,incr1,grid1);
end
vals2=dlmread([IntData,'IdealCube_vals.csv']);

names=caseread([DataColl,filename]);
[Nn,~]=size(names);
```

```
DataOut=zeros(Nn*N,27);
rt=zeros(3,N);

p=1;

for a=1:Nn
    table=dlmread([DataColl,names(a,:)]);

    rt1=table(1:N,1:3)';
    qt=table(1:N,28:31)';

    rSeed=rt1(:,1);

    %shift the median of the position into the center of the cube
    %so that we get SOME good points out of each data set.  We need
    %to do this because we gathered the data on two differenct days.
    %All the data collected at the same time does not have this
    %problem.  We use the median to limit the effect of a few bad
    %values.
    if (N>= 1000)
        K=1000;
    else
        K=N;
    end
    rc1=median(rt1(:,1:K),2);
    delta=rc1-rc;
    for n=1:N
        rt(:,n)=rt1(:,n)-delta;
    end

    for n=1:N

        %get the sensor rotation matrix
        qt(:,n)=qt(:,n)/sqrt(qt(:,n)'*qt(:,n));
        As=AerToA(QtoAer(qt(:,n)));

        %generate the measured field
        svec=GetField(rt(:,n),lut,min1,incr1,grid1);
        snoise=random('norm',snoise_mean,s_stdev);
```

```
svec=svec+snoise;
smeas=[svec(1:3)';svec(4:6)';svec(7:9)'];
s4meas=As'*smeas;
s4vec_meas=[s4meas(1,:),s4meas(2,:),s4meas(3,:)]';
rs=rFindR(rSeed,s4vec_meas,xyz,vals');

%Check if solution is inside of interpolation volume
flag=0;
for j=1:3
    if (rs(j)<xyz((j-1)*2+1))
        flag=1;
    end

    if (rs(j)>xyz((j-1)*2+2))
        flag=1;
    end
end

if (flag==0)

    %keep the answers since we inside the interpolation volume
    rest=rs;

    %generate the estimated signal matrix
    sest_vec=Interp(rest(:,1),xyz,vals')';
    sest=[sest_vec(1:3)';sest_vec(4:6)';sest_vec(7:9)'];
    qest=SolveForOrientation(sest,s4meas);
    qest=qest/sqrt(qest'*qest);

    %generate the S4vec error
    aerEst1=QtoAer(qest);
    Aest=AerToA(aerEst1);
    s4_err=sest-Aest*s4meas;
    s4vec_err=[s4_err(1,:),s4_err(2,:),s4_err(3,:)]';

    %track PnO errors
    r_err=rest-rt(:,n);
    rT_err=sqrt(r_err'*r_err);
    aer_err=QtoAer(Qmul(qest,qt(:,n),0,1))';
```

```
                    aerT_err=sqrt(aer_err'*aer_err);
                    serr=sqrt(s4vec_err'*s4vec_err);

                    DataOut(p,:)=[a,rT_err,aerT_err,serr,s4vec_err',
                             rt(:,n)',qt(:,n)',rest',qest'];
                    p=p+1;
               end
               rSeed=rs;

          end %for n=
      end%for a=

      dlmwrite([IntData,prefix,'_stats.csv'],DataOut(1:p-1,:));
      rM=median(DataOut(1:p-1,2));
      aerM=median(DataOut(1:p-1,3));

end
```

C.10 RUNMAPSTATS(·)

```
function    RunMapStats(SimDir)
%a function to run statsitics on the map.
% SimDir    root directory of data files

%set paths
DataColl=cat(2,SimDir,'DataCollection\');
IntData=cat(2,SimDir,'InterpolationData\');
LUTdata=cat(2,SimDir,'LUTdata\');

%read in the sampled map and its parameters
lutS=dlmread([DataColl,'SampledLut.csv']);
paramS=dlmread([DataColl,'SampleMapParam.csv']);
minS=paramS(:,1);
incrS=paramS(:,2);
gridS=paramS(:,3);
maxS=paramS(:,4);
MapStructS=CreateMapStruct(lutS,minS,incrS,gridS);
Ns=gridS(1)*gridS(2)*gridS(3);

%bring in the lut and lut parameters
```

```
lutM=dlmread(horzcat(SimDir,'LUTdata\LCDmap_lut.csv'));
paramM=dlmread([LUTdata,'LCDmap_param.csv']);
minM=paramM(:,1);
incrM=paramM(:,2);
gridM=paramM(:,3);
%maxM=minM+incrM.*(gridM-1);
MapStructM=CreateMapStruct(lutM,minM,incrM,gridM);

%Create the polynomial correction
%the polynomial runs off a grid data set of measurements taken in the
%mapped volume.  This data does not have to be gridded but does require
%mechanical registration similar to the mapper
n=1;
r_tru=zeros(3,Ns);
r_dp=zeros(3,Ns);
dr=zeros(3,Ns);

q_tru=zeros(4,Ns);
q_dp=zeros(4,Ns);
dq=zeros(4,Ns);

for i=1:gridS(1)
    for j=1:gridS(2);
        for k=1:gridS(3);
            r_tru(:,n) = minS+([i;j;k]-1).*incrS;
            q_tru(:,n) = [1;0;0;0];

            %now get the measured data for this point
            svm=GetField(r_tru(:,n),MapStructM);
            sm=[svm(1:3,1)';svm(4:6,1)';svm(7:9,1)'];

            %solve using the dipole algorithm
            pno = DipolePnO(sm,r_tru(:,n));
            r_dp(:,n)=pno(1:3);
            q_dp(:,n)=AerToQ(pno(4:6));
            q_dp(:,n)= NormQ(q_dp(:,n));

            %calculate the deltas
            dr(:,n)=r_dp(:,n)-r_tru(:,n);
```

```
                        dq(:,n)=Qmul(q_dp(:,n),q_tru(:,n),0,1);
                        n=n+1;
                    end
            end
end
N=n-1;

%Now create the polynomial
p1_2 = polyfitn(r_dp',dr(1,:)',2);
p2_2 = polyfitn(r_dp',dr(2,:)',2);
p3_2 = polyfitn(r_dp',dr(3,:)',2);
p4_2 = polyfitn(r_dp',dq(1,:)',2);
p5_2 = polyfitn(r_dp',dq(2,:)',2);
p6_2 = polyfitn(r_dp',dq(3,:)',2);
p7_2 = polyfitn(r_dp',dq(4,:)',2);

%Now create the polynomial
p1_3 = polyfitn(r_dp',dr(1,:)',3);
p2_3 = polyfitn(r_dp',dr(2,:)',3);
p3_3 = polyfitn(r_dp',dr(3,:)',3);
p4_3 = polyfitn(r_dp',dq(1,:)',3);
p5_3 = polyfitn(r_dp',dq(2,:)',3);
p6_3 = polyfitn(r_dp',dq(3,:)',3);
p7_3 = polyfitn(r_dp',dq(4,:)',3);

%Now create the polynomial
p1_4 = polyfitn(r_dp',dr(1,:)',4);
p2_4 = polyfitn(r_dp',dr(2,:)',4);
p3_4 = polyfitn(r_dp',dr(3,:)',4);
p4_4 = polyfitn(r_dp',dq(1,:)',4);
p5_4 = polyfitn(r_dp',dq(2,:)',4);
p6_4 = polyfitn(r_dp',dq(3,:)',4);
p7_4 = polyfitn(r_dp',dq(4,:)',4);

%create the grid points for the sampled map
rGridS=zeros(Ns,3);
n=1;
lutSa=zeros(Ns,9);
ijk=zeros(3,Ns);
```

```
for i=1:gridS(1)
    for j=1:gridS(2)
        for k=1:gridS(3)
            ijk(:,n)=[i j k]';
            rGridS(n,:)=minS+incrS.*(ijk(:,n)-1);
            lutSa(n,:)=Lookup(rGridS(n,:)',MapStructM)';
            n=n+1;
        end
    end
end


%Repeat number 2 with random positions to evaluate overall performance
%here we solve for random position/orientation inside the sampled table,
%using the master signal matrix for measured field data

Nr=1000;
r3(:,1)=minS(1)+(maxS(1)-minS(1))*rand(Nr,1);
r3(:,2)=minS(2)+(maxS(2)-minS(2))*rand(Nr,1);
r3(:,3)=minS(3)+(maxS(3)-minS(3))*rand(Nr,1);
aer3(:,1)=-pi+2*pi*rand(Nr,1);
aer3(:,2)=-pi/2+pi*rand(Nr,1);
aer3(:,3)=-pi+2*pi*rand(Nr,1);

%get polynomial corrections (2)
drErr2=zeros(3,Nr);
dqErr2=zeros(4,Nr);
drErr2(1,:)=polyvaln(p1_2,r3)';
drErr2(2,:)=polyvaln(p2_2,r3)';
drErr2(3,:)=polyvaln(p3_2,r3)';
dqErr2(1,:)=polyvaln(p4_2,r3)';
dqErr2(2,:)=polyvaln(p5_2,r3)';
dqErr2(3,:)=polyvaln(p6_2,r3)';
dqErr2(4,:)=polyvaln(p7_2,r3)';

%get polynomial corrections (3)
drErr3=zeros(3,Nr);
dqErr3=zeros(4,Nr);
drErr3(1,:)=polyvaln(p1_3,r3)';
drErr3(2,:)=polyvaln(p2_3,r3)';
```

```
drErr3(3,:)=polyvaln(p3_3,r3)';
dqErr3(1,:)=polyvaln(p4_3,r3)';
dqErr3(2,:)=polyvaln(p5_3,r3)';
dqErr3(3,:)=polyvaln(p6_3,r3)';
dqErr3(4,:)=polyvaln(p7_3,r3)';

%get polynomial corrections (4)
drErr4=zeros(3,Nr);
dqErr4=zeros(4,Nr);
drErr4(1,:)=polyvaln(p1_4,r3)';
drErr4(2,:)=polyvaln(p2_4,r3)';
drErr4(3,:)=polyvaln(p3_4,r3)';
dqErr4(1,:)=polyvaln(p4_4,r3)';
dqErr4(2,:)=polyvaln(p5_4,r3)';
dqErr4(3,:)=polyvaln(p6_4,r3)';
dqErr4(4,:)=polyvaln(p7_4,r3)';

for n=1:Nr

    %construct the measured field
    rtst=r3(n,:)';
    sv3=GetField(rtst,MapStructM);
    sm=[sv3(1:3)';sv3(4:6)';sv3(7:9)'];
    A3=AerToA(aer3(n,:)');
    q3=AerToQ(aer3(n,:)');
    sm3=A3'*sm;

    %solve with dipole algorithm
    pno = DipolePnO(sm3,rtst);
    rest_dp=pno(1:3);
    rErr_dp=rest_dp-rtst;
    rErrT_dp=sqrt(rErr_dp'*rErr_dp);
    qest_dp=AerToQ(pno(4:6));
    qest_dp=NormQ(qest_dp);
    qerr_dp=Qmul(qest_dp,q3,0,1);
    aerErr_dp=QtoAer(qerr_dp)';
    aerErrT_dp=sqrt(aerErr_dp'*aerErr_dp);
    out_dp(n,:)=[rtst;rest_dp;rErr_dp;rErrT_dp;aerErr_dp;aerErrT_dp]';
```

```
%correct pno solution with polynomial (2)
rest_p=rest_dp-drErr2(:,n);        %subtract predicted r error
rErr_p=rest_p-rtst;
rErrT_p=sqrt(rErr_p'*rErr_p);
dq=NormQ(dqErr2(:,n));
qest_p=Qmul(dq,qest_dp,1,0);       %backout predicted q error
qerr_p=Qmul(qest_p,q3,0,1);
aerErr_p=QtoAer(qerr_p)';
aerErrT_p=sqrt(aerErr_p'*aerErr_p);
out_p2(n,:)=[rtst;rest_p;rErr_p;rErrT_p;aerErr_p;aerErrT_p]';

%correct pno solution with polynomial (3)
rest_p=rest_dp-drErr3(:,n);        %subtract predicted r error
rErr_p=rest_p-rtst;
rErrT_p=sqrt(rErr_p'*rErr_p);
dq=NormQ(dqErr3(:,n));
qest_p=Qmul(dq,qest_dp,1,0);       %backout predicted q error
qerr_p=Qmul(qest_p,q3,0,1);
aerErr_p=QtoAer(qerr_p)';
aerErrT_p=sqrt(aerErr_p'*aerErr_p);
out_p3(n,:)=[rtst;rest_p;rErr_p;rErrT_p;aerErr_p;aerErrT_p]';

%correct pno solution with polynomial (4)
rest_p=rest_dp-drErr4(:,n);        %subtract predicted r error
rErr_p=rest_p-rtst;
rErrT_p=sqrt(rErr_p'*rErr_p);
dq=NormQ(dqErr4(:,n));
qest_p=Qmul(dq,qest_dp,1,0);       %backout predicted q error
qerr_p=Qmul(qest_p,q3,0,1);
aerErr_p=QtoAer(qerr_p)';
aerErrT_p=sqrt(aerErr_p'*aerErr_p);
out_p4(n,:)=[rtst;rest_p;rErr_p;rErrT_p;aerErr_p;aerErrT_p]';

%solve for PnO with stick map compensation
rest3=SolveStS(rtst,MapStructS,sm3);
rErr3=rest3-rtst;
rErrT3=sqrt(rErr3'*rErr3);
svec_est3=GetField(rest3,MapStructS);
sest3=[svec_est3(1:3)';svec_est3(4:6)';svec_est3(7:9)'];
```

```
       A3=(sm3/sest3)';
       A3=Orthog(A3);
       qest3=AerToQ(AtoAer(A3));
       qest3=qest3/sqrt(qest3'*qest3);
       qerr3=Qmul(qest3,q3,0,1);
       aerErr3=QtoAer(qerr3)';
       aerErrT3=sqrt(aerErr3'*aerErr3);
       out3(n,:)=[rtst;rest3;rErr3;rErrT3;aerErr3;aerErrT3]';


end
dlmwrite([DataColl,'Stats_Random_dp.csv'],out_dp);
dlmwrite([DataColl,'Stats_Random_p2.csv'],out_p2);
dlmwrite([DataColl,'Stats_Random_p3.csv'],out_p3);
dlmwrite([DataColl,'Stats_Random_p4.csv'],out_p4);
dlmwrite([DataColl,'Stats_Random_stick.csv'],out3);


end
```

C.11 RUNMAPSTATS_MULTILEVEL(·)

```
function   RunMapStats_MultiLevel(suffix,dist)
% a funciton to run statistical anlysis of the results
% suffix        suffix to apply to output files
% dist          artificial distorion level

%set paths
DataColl=cat(2,SimDir,'DataCollection\');
IntData=cat(2,SimDir,'InterpolationData\');
LUTdata=cat(2,SimDir,'LUTdata\');

%read in the sampled map and its parameters
lutS=dlmread([DataColl,'SampledLut.csv']);
lutS=lutS*dist; %scale distortion
paramS=dlmread([DataColl,'SampleMapParam.csv']);
minS=paramS(:,1);
incrS=paramS(:,2);
gridS=paramS(:,3);
maxS=paramS(:,4);
MapStructS=CreateMapStruct(lutS,minS,incrS,gridS);
Ns=gridS(1)*gridS(2)*gridS(3);
```

```
%bring in the lut and lut parameters
lutM=dlmread(horzcat(MatLabSim,'LUTdata\LCDmap_lut.csv'));
lutM=dist*lutM; %scale distortion
paramM=dlmread([LUTdata,'LCDmap_param.csv']);
minM=paramM(:,1);
incrM=paramM(:,2);
gridM=paramM(:,3);
%maxM=minM+incrM.*(gridM-1);
MapStructM=CreateMapStruct(lutM,minM,incrM,gridM);

%Create the polynomial correction
%the polynomial runs off a grid data set of measurements taken in the
%mapped volume.  This data does not have to be gridded but does require
%mechanical registration similar to the mapper
n=1;
r_tru=zeros(3,Ns);
r_dp=zeros(3,Ns);
dr=zeros(3,Ns);

q_tru=zeros(4,Ns);
q_dp=zeros(4,Ns);
dq=zeros(4,Ns);

for i=1:gridS(1)
    for j=1:gridS(2);
        for k=1:gridS(3);
            r_tru(:,n) = minS+([i;j;k]-1).*incrS;
            q_tru(:,n) = [1;0;0;0];

            %now get the measured data for this point
            svm=GetField(r_tru(:,n),MapStructM);
            sm=[svm(1:3,1)';svm(4:6,1)';svm(7:9,1)'];

            %solve using the dipole algorithm
            pno = DipolePnO(sm,r_tru(:,n));
            r_dp(:,n)=pno(1:3);
            q_dp(:,n)=AerToQ(pno(4:6));
            q_dp(:,n)= NormQ(q_dp(:,n));
```

```
            %calculate the deltas
            dr(:,n)=r_dp(:,n)-r_tru(:,n);
            dq(:,n)=Qmul(q_dp(:,n),q_tru(:,n),0,1);
            n=n+1;
        end
    end
end
N=n-1;

%Now create the polynomial
p1_2 = polyfitn(r_dp',dr(1,:)',2);
p2_2 = polyfitn(r_dp',dr(2,:)',2);
p3_2 = polyfitn(r_dp',dr(3,:)',2);
p4_2 = polyfitn(r_dp',dq(1,:)',2);
p5_2 = polyfitn(r_dp',dq(2,:)',2);
p6_2 = polyfitn(r_dp',dq(3,:)',2);
p7_2 = polyfitn(r_dp',dq(4,:)',2);

%Now create the polynomial
p1_3 = polyfitn(r_dp',dr(1,:)',3);
p2_3 = polyfitn(r_dp',dr(2,:)',3);
p3_3 = polyfitn(r_dp',dr(3,:)',3);
p4_3 = polyfitn(r_dp',dq(1,:)',3);
p5_3 = polyfitn(r_dp',dq(2,:)',3);
p6_3 = polyfitn(r_dp',dq(3,:)',3);
p7_3 = polyfitn(r_dp',dq(4,:)',3);

%Now create the polynomial
p1_4 = polyfitn(r_dp',dr(1,:)',4);
p2_4 = polyfitn(r_dp',dr(2,:)',4);
p3_4 = polyfitn(r_dp',dr(3,:)',4);
p4_4 = polyfitn(r_dp',dq(1,:)',4);
p5_4 = polyfitn(r_dp',dq(2,:)',4);
p6_4 = polyfitn(r_dp',dq(3,:)',4);
p7_4 = polyfitn(r_dp',dq(4,:)',4);

%create the grid points for the sampled map
rGridS=zeros(Ns,3);
```

```
n=1;
ijk=zeros(3,Ns);
for i=1:gridS(1)
    for j=1:gridS(2)
        for k=1:gridS(3)
            ijk(:,n)=[i j k]';
            rGridS(n,:)=minS+incrS.*(ijk(:,n)-1);
            n=n+1;
        end
    end
end

Nr=1000;
r3(:,1)=minS(1)+(maxS(1)-minS(1))*rand(Nr,1);
r3(:,2)=minS(2)+(maxS(2)-minS(2))*rand(Nr,1);
r3(:,3)=minS(3)+(maxS(3)-minS(3))*rand(Nr,1);

dr3(:,1)=-0.002+0.004*rand(Nr,1);
dr3(:,2)=-0.002+0.004*rand(Nr,1);
dr3(:,3)=-0.002+0.004*rand(Nr,1);

aer3(:,1)=-pi+2*pi*rand(Nr,1);
aer3(:,2)=-pi/2+pi*rand(Nr,1);
aer3(:,3)=-pi+2*pi*rand(Nr,1);

%get polynomial corrections (2)
drErr2=zeros(3,Nr);
dqErr2=zeros(4,Nr);
drErr2(1,:)=polyvaln(p1_2,r3)';
drErr2(2,:)=polyvaln(p2_2,r3)';
drErr2(3,:)=polyvaln(p3_2,r3)';
dqErr2(1,:)=polyvaln(p4_2,r3)';
dqErr2(2,:)=polyvaln(p5_2,r3)';
dqErr2(3,:)=polyvaln(p6_2,r3)';
dqErr2(4,:)=polyvaln(p7_2,r3)';

%get polynomial corrections (3)
drErr3=zeros(3,Nr);
dqErr3=zeros(4,Nr);
```

```
drErr3(1,:)=polyvaln(p1_3,r3)';
drErr3(2,:)=polyvaln(p2_3,r3)';
drErr3(3,:)=polyvaln(p3_3,r3)';
dqErr3(1,:)=polyvaln(p4_3,r3)';
dqErr3(2,:)=polyvaln(p5_3,r3)';
dqErr3(3,:)=polyvaln(p6_3,r3)';
dqErr3(4,:)=polyvaln(p7_3,r3)';

%get polynomial corrections (4)
drErr4=zeros(3,Nr);
dqErr4=zeros(4,Nr);
drErr4(1,:)=polyvaln(p1_4,r3)';
drErr4(2,:)=polyvaln(p2_4,r3)';
drErr4(3,:)=polyvaln(p3_4,r3)';
dqErr4(1,:)=polyvaln(p4_4,r3)';
dqErr4(2,:)=polyvaln(p5_4,r3)';
dqErr4(3,:)=polyvaln(p6_4,r3)';
dqErr4(4,:)=polyvaln(p7_4,r3)';

out3=zeros(Nr,14);
out_dp=zeros(Nr,14);
out_p2=zeros(Nr,14);
out_p3=zeros(Nr,14);
out_p4=zeros(Nr,14);

for n=1:Nr

    %construct the measured field
    rtru=r3(n,:)';
    rseed=rtru+dr3(n,:)';
    g=dist*Lookup(rtru,MapStructM);
    sv3=GetField(rtru,MapStructM);
    sm=[sv3(1:3)';sv3(4:6)';sv3(7:9)'];
    A3=AerToA(aer3(n,:)');
    q3=AerToQ(aer3(n,:)');
    sm3=A3'*sm;

    %solve with dipole algorithm
    pno = DipolePn0(sm3,rseed);
```

```
rest_dp=pno(1:3);
rErr_dp=rest_dp-rtru;
rErrT_dp=sqrt(rErr_dp'*rErr_dp);
qest_dp=AerToQ(pno(4:6));
qest_dp=NormQ(qest_dp);
qerr_dp=Qmul(qest_dp,q3,0,1);
aerErr_dp=QtoAer(qerr_dp)';
aerErrT_dp=sqrt(aerErr_dp'*aerErr_dp);
out_dp(n,:)=[rtru;rest_dp;rErr_dp;rErrT_dp;aerErr_dp;aerErrT_dp]';

%correct pno solution with polynomial (2)
rest_p=rest_dp-drErr2(:,n);        %subtract predicted r error
rErr_p=rest_p-rtru;
rErrT_p=sqrt(rErr_p'*rErr_p);
dq=NormQ(dqErr2(:,n));
qest_p=Qmul(dq,qest_dp,1,0);     %backout predicted q error
qerr_p=Qmul(qest_p,q3,0,1);
aerErr_p=QtoAer(qerr_p)';
aerErrT_p=sqrt(aerErr_p'*aerErr_p);
out_p2(n,:)=[rtru;rest_p;rErr_p;rErrT_p;aerErr_p;aerErrT_p]';

%correct pno solution with polynomial (3)
rest_p=rest_dp-drErr3(:,n);        %subtract predicted r error
rErr_p=rest_p-rtru;
rErrT_p=sqrt(rErr_p'*rErr_p);
dq=NormQ(dqErr3(:,n));
qest_p=Qmul(dq,qest_dp,1,0);     %backout predicted q error
qerr_p=Qmul(qest_p,q3,0,1);
aerErr_p=QtoAer(qerr_p)';
aerErrT_p=sqrt(aerErr_p'*aerErr_p);
out_p3(n,:)=[rtru;rest_p;rErr_p;rErrT_p;aerErr_p;aerErrT_p]';

%correct pno solution with polynomial (4)
rest_p=rest_dp-drErr4(:,n);        %subtract predicted r error
rErr_p=rest_p-rtru;
rErrT_p=sqrt(rErr_p'*rErr_p);
dq=NormQ(dqErr4(:,n));
qest_p=Qmul(dq,qest_dp,1,0);     %backout predicted q error
qerr_p=Qmul(qest_p,q3,0,1);
```

```
        aerErr_p=QtoAer(qerr_p)';
        aerErrT_p=sqrt(aerErr_p'*aerErr_p);
        out_p4(n,:)=[rtru;rest_p;rErr_p;rErrT_p;aerErr_p;aerErrT_p]';

        %solve for PnO with stick map compensation
        rest3=SolveStS(rseed,MapStructS,sm3);
        rErr3=rest3-rtru;
        rErrT3=sqrt(rErr3'*rErr3);
        svec_est3=GetField(rest3,MapStructS);
        sest3=[svec_est3(1:3)';svec_est3(4:6)';svec_est3(7:9)'];
        A3=(sm3/sest3)';
        A3=Orthog(A3);
        qest3=AerToQ(AtoAer(A3));
        qest3=qest3/sqrt(qest3'*qest3);
        qerr3=Qmul(qest3,q3,0,1);
        aerErr3=QtoAer(qerr3)';
        aerErrT3=sqrt(aerErr3'*aerErr3);
        out3(n,:)=[rtru;rest3;rErr3;rErrT3;aerErr3;aerErrT3]';

end
dlmwrite([DataColl,'Stats_Random_dp_',suffix,'.csv'],out_dp);
dlmwrite([DataColl,'Stats_Random_p2_',suffix,'.csv'],out_p2);
dlmwrite([DataColl,'Stats_Random_p3_',suffix,'.csv'],out_p3);
dlmwrite([DataColl,'Stats_Random_p4_',suffix,'.csv'],out_p4);
dlmwrite([DataColl,'Stats_Random_stick_',suffix,'.csv'],out3);

end
```

C.12 SORTTABLEBYVAL(·)

A function that sorts a data table by value.

```
function tOut=SortTableByVal(tIn,col,val,dir)

[N,M]=size(tIn);
t=zeros(N,M);

m=1;
for n=1:N
    if (((dir==1)&(tIn(n,col)>val))|((dir==0)&(tIn(n,col)<val)))
        t(m,:)=tIn(n,:);
```

```
        m=m+1;
    end
end
tOut=t(1:m-1,:);
```

APPENDIX D

MatLab Library

These functions are shared among the three experiments (DQEKF, MMDQ, and IVC).

D.1 ATOAER(·)

A function to convert a rotation matrix (A) to Euler angles (AER).

```
function aer = AtoAer(A)
%AtoAer - convert a rotation matrix to Euler angles
%convert a rotation matrix to Euler angles in radians

psi=atan2(A(2,1),A(1,1));
s_theta=-A(3,1);
c_theta=cos(psi)*A(1,1)+sin(psi)*A(2,1);
theta=atan2(s_theta,c_theta);
s_phi=sin(psi)*A(1,3)-cos(psi)*A(2,3);
c_phi=-sin(psi)*A(1,2)+cos(psi)*A(2,2);
phi=atan2(s_phi,c_phi);
aer(1)=psi;
aer(2)=theta;
aer(3)=phi;
```

D.2 AERTOA(·)

A function to convert Euler angles (AER) to a rotation matrix (A).

```
function A = AerToA(Aer)
%Convert an Euler Angle set to a rotation matrix
%Convert a Euler Angle (in radians) to a rotation matrix

A=eye(3);
psi=Aer(1);
theta=Aer(2);
phi=Aer(3);
```

```
c_psi=cos(psi);
s_psi=sin(psi);
c_theta=cos(theta);
s_theta=sin(theta);
c_phi=cos(phi);
s_phi=sin(phi);

A(1,1)= c_psi*c_theta;
A(2,1)= s_psi*c_theta;
A(3,1)=-s_theta;
A(1,2)=-s_psi*c_phi+c_psi*s_theta*s_phi;
A(2,2)= c_psi*c_phi+s_psi*s_theta*s_phi;
A(3,2)= c_theta*s_phi;
A(1,3)= s_psi*s_phi+c_psi*s_theta*c_phi;
A(2,3)=-c_psi*s_phi+s_psi*s_theta*c_phi;
A(3,3)= c_theta*c_phi;
```

D.3   AERTOQ(·)

A function to convert Euler angles (AER) to quaternions (Q).

```
function q = AerToQ(aer)
%AerToQ: convert euler angles to quaternion orientation
%Convert Euler angles (radians) to normalized quaternions

q=zeros(4,1);
psi=aer(1);
theta=aer(2);
phi=aer(3);

c_psi=cos(psi/2);
c_theta=cos(theta/2);
c_phi=cos(phi/2);
s_psi=sin(psi/2);
s_theta=sin(theta/2);
s_phi=sin(phi/2);

q(1)=c_psi*c_theta*c_phi+s_psi*s_theta*s_phi;
q(2)=c_psi*c_theta*s_phi-s_psi*s_theta*c_phi;
q(3)=c_psi*s_theta*c_phi+s_psi*c_theta*s_phi;
q(4)=s_psi*c_theta*c_phi-c_psi*s_theta*s_phi;
```

D.4 CREATELMM_PARAMSTRUCT(·)

Create a structure for the Minpac optimization (LMM).

```
function param =...
    CreateLMM_ParamStruct(a,x,ia,sig,ConvThresh,ConvLimit,IterLimit)
%Create  parameter struvture for the LMM solver
% a       seed
% x       data points
% ia      flags to make associated a entries constant
% sig
% ConvThresh   quit if error is below this value
% ConvLimit    number of iterations at same value until quit
% IterLimit    total iteration limit

        [ma,~]=size(a);                 %number of independent variables

        param.a=a;
        param.atry=zeros(ma);
        param.alpha=zeros(ma,ma);
        param.beta=zeros(ma);
        param.alamda=0.001;
        param.iter=0;                   %iterations to convergence or quit
        param.ia=ia;                    %indicates fixed "a" entries
        param.mfit=0;
        param.y=x;
        param.sig=sig;

        param.ConvLimit=ConvLimit;      %convergene iteration limit
        param.ConvThresh=ConvThresh;    %convergence threshold
        param.IterLimit=IterLimit;      %iteration limit

        param.chisq=10e+20;             %current error
end
```

D.5 DIPOLEPNO(·)

Compute the PnO from a dipole field matrix.

```
function pno = DipolePnO(S,seed)
%DipolePnO() - compute the dipole PnO
```

```
% S - normalized signal matrix
%seed    a position near the solution
%
%returns a vector with position in meters
%and Euler orientation in radians

sts=S'*S;
r6=6/(sts(1,1)+sts(2,2)+sts(3,3));
rm=10^(log10(r6)/6);
R=r6*sts-eye(3);
R=R/3;
r=seed;
for i=1:10
    r=R*r;
end
ru=r/sqrt(r'*r);
r=rm*ru;

%get orientation
s=GetIdealField(r);
At=S/s;
A=Orthog(At');
aer=AtoAer(A);
pno=[r; aer'];
```

D.6   GAUSSJ(·)

This function was taken from *Numerical Recipes in C++* by W. Press, W. Vetterling, S. Teukolsky, and B. Flannery.

```
function [a,b] = gaussj(a,b)
%gaussian-Jordan elimination
%    returns a    inverse of a
%             b    corresponding solution vectors

%*** This function taken from "Numerical Recipes for C" ***

[n,~]=size(a);
[~,m]=size(b);
indxr=zeros(n);      %declare indxr array
indxc=zeros(n);      %declare indxc array
```

```
ipiv=zeros(n);          %clear pivot flags
icol=1;
irow=1;

for i=1:n
    % This is the main loop over the columns to be reduced
    big=0;

    %find the pivot element
    for j=1:n
        % this is the outer loop of the search for a pivot element
        if (ipiv(j) ~=1)
            for k=1:n
                if (ipiv(k)==0)
                    if (abs(a(j,k))>=big)
                        big=abs(a(j,k));
                        irow=j;
                        icol=k;
                    end
                end
            end
        end
    end

    ipiv(icol)=ipiv(icol)+1;

    %we now have the pivot element so we interchange rows (if needed)
    %to put the pivot element on the diagonal.  The columns are not
    %physically interchanged, only relabeled; indxc(i), the column of
    %the (i+1)th piovot element, is the (i+1)th column that is reduced,
    %while indexr(i) is the row in which that pivot element was originally
    %located.  If indxr(i) .ne indxc(i) there is an implied column
    %interchange.  With this form of bookkeeping the solution
    %b's will end up in the correct order and the inverse matrix will
    %be scrambled by columns

    if (irow ~=icol)
        for l=1:n
            temp=a(irow,l);
```

```matlab
            a(irow,l)=a(icol,l);
            a(icol,l)=temp;
        end

        for l=1:m
            temp=b(irow,l);
            b(irow,l)=b(icol,l);
            b(icol,l)=temp;
        end
    end

    %we are now resdy to divide the pivot row by the pivot element,
    %located at irow, icol
    indxr(i)=irow;
    indxc(i)=icol;
    %if (a(icol,icol)==0) nerror("gaussj: Singular Matrix");
    pivinv=1/a(icol,icol);
    a(icol,icol)=1;
    for l=1:n
        a(icol,l)=a(icol,l)*pivinv;
    end
    for l=1:m
        b(icol,l)=b(icol,l)*pivinv;
    end

    %now reduce the other rows, except for pivot
    for ll=1:n
        if (ll~=icol)
            dum=a(ll,icol);
            a(ll,icol)=0;
            for l=1:n
                a(ll,l)=a(ll,l)-a(icol,l)*dum;
            end
            for l=1:m
                b(ll,l)=b(ll,l)-b(icol,l)*dum;
            end
        end
    end
end
```

```
%this is the end of the main loop over columns aof the reduction.   It
%only remains to unscramble the solution in view of the column
%interchanges.  We do this by interchanging pairs of columns in the
%reverse order that the permutation was built up.
l=n;
while (l>0)
    if (indxr(l) ~= indxc(l))
        for k=1:n
            temp=a(k,indxr(l));
            a(k,indxr(l))=a(k,indxc(l));
            a(k,indxc(l))=temp;
        end
    end
    l=l-1;
end

end
```

D.7   GETBTB(·)

```
function sts =GetBtB(s4vec)
%GetBtB() - compute the StS product of a signal matrix
s4=[s4vec(1:3)';s4vec(4:6)';s4vec(7:9)'];
sts1=s4'*s4;
sts=[sts1(1,1) sts1(1,2) sts1(1,3) sts1(2,2) sts1(2,3) sts1(3,3)];
```

D.8   GETDQ(·)

```
function    dq=GetDq(x,t)
%compute the delta quaternion for a angular velocity (no accelration)
% x  series of quaternion vectors
% t  time base in seconds

    w=x(1:3,1);
    W=sqrt(w'*w);
    if (W==0)
        W=1e-15;
    end
    theta=W*t/2;
```

```
    dq(1,1)=cos(theta);
    dq(2:4,1)=w*2*sin(theta)/W;
    dq=dq/sqrt(dq'*dq);

    %force a position rotation
    if (dq(1)<0)
        dq=-dq;
    end
```

D.9   GETIDEALFIELD(·)

A function to compute the ideal normalized 3D dipole field.

```
function S = GetIdealField(r1)
%Generate the signal matrix for a dipole field
% This function generates the normalized unrotated
%signal matrix that would be generated at field
%point r in a dipole field.

r=r1(1:3);

B=zeros(3);
R2=r'*r;
R=sqrt(R2);
R3=R2*R;
R5=R2*R2*R;

S(1,1)=(3*r(1)*r(1)/R2-1)/R3;
S(1,2)=3*r(1)*r(2)/R5;
S(1,3)=3*r(1)*r(3)/R5;
S(2,1)=S(1,2);
S(2,2)=(3*r(2)*r(2)/R2-1)/R3;
S(2,3)=3*r(2)*r(3)/R5;
S(3,1)=S(1,3);
S(3,2)=S(2,3);
S(3,3)=(3*r(3)*r(3)/R2-1)/R3;
```

D.10 GETFIELD(·)

A function to get the field of a known point using a LUT.

```
function s4vec = GetField(rIn,MapStruct)
%find the measured field of a point in a map
%rIn          position
%MapStruct    LUT map structure
%
r=rIn(1:3,1);
gvec=Lookup(r,MapStruct)';
g4=[gvec(1:3)';gvec(4:6)';gvec(7:9)'];
s4=GetIdealField(r);
s4=s4+g4;
s4vec=[s4(1,:),s4(2,:),s4(3,:)]';
```

D.11 INTERP(·)

```
function ival = Interp(r,xyz,vals)
%Interp() - Trilinear interpolation function for LMM PnO Solver
% r      target position
% xyz    indices of associated points from table
% vals   values from table for each indice

tmp=[0 0 0 0 0 0 0 0];
ival=[0 0 0 0 0 0 0 0];

for a=1:4
    for b=1:9
        v0=vals(a,b);
        v1=vals(a+4,b);
        s0=xyz(1);
        s1=xyz(2);
        s=r(1);
        tmp(a,b)=v0+(s-s0)*(v1-v0)/(s1-s0);
    end
end

for a=1:2
    for b=1:9
        v0=tmp(a,b);
```

```
        v1=tmp(a+2,b);
        s0=xyz(3);
        s1=xyz(4);
        s=r(2);
        tmp(a+4,b)=v0+(s-s0)*(v1-v0)/(s1-s0);
    end
end

for b=1:9
    v0=tmp(5,b);
    v1=tmp(6,b);
    s0=xyz(5);
    s1=xyz(6);
    s=r(3);
    ival(b)=v0+(s-s0)*(v1-v0)/(s1-s0);
end
```

D.12 LOOKUP(·)

```
function bvec = Lookup(rIn,MapStruct)
%Lookup() - Lookup a field value from a LMM map
%rIn           target position
%MapStruct     LUT mapping structure

r=rIn(1:3,1);

ijk=GetCornerIJK(r,MapStruct);
xyz=GetCornerXYZ(ijk,MapStruct);
vals=GetCornerVals(ijk,MapStruct);
bvec=Interp(r,xyz,vals);
```

D.13 MRQCOEF(·)

This function was taken from *Numerical Recipes in C++* by W. Press, W. Vetterling, S. Teukolsky, and B. Flannery.

```
function [alpha,beta,chisq] = mrqcof(param,a,alpha,beta,func )
%This is an internal function (private) of the LMM solver

%*** modified from "Numerical Methods in C" ***
```

```
%create local copy of parameters
ma=size(a);
npt=size(param.y);

%clear curvature matrix
for j=0:(param.mfit-1)
    for k=0:j
        alpha(j+1,k+1)=0;
    end
    beta(j+1)=0;
end

%get the function values and partial derivatives at the test point
[ymod,dyda]=func(param.y,a);

chisq=0;
for i=0:(npt-1)
    sig2i=1.0/param.sig(i+1)^2;
    dy=param.y(i+1)-ymod(i+1);
    j=0;
    for l=0:(ma-1)
        if (param.ia(l+1)>0)
            wt=dyda(i+1,l+1)*sig2i;
            k=0;
            for m=0:l
                if (param.ia(m+1)>0)
                    alpha(j+1,k+1)=alpha(j+1,k+1)+wt*dyda(i+1,m+1);
                    k=k+1;
                end
            end
            beta(j+1)=beta(j+1)+dy*wt;
            j=j+1;
        end
    end

    %find Chi-squared
    chisq = chisq+dy^2*sig2i;
end
```

```
        %fill in the symetric side
        for j=0:(param.mfit-1)
            for k=0:(j-1)
                alpha(k+1,j+1)=alpha(j+1,k+1);
            end
        end

end
```

D.14 MRQMIN(·)

This function was taken from *Numerical Recipes in C++* by W. Press, W. Vetterling, S. Teukolsky, and B. Flannery.

```
function param = mrqmin( param,func )
% this function runs a single pass of the LMM optimization
% *** modified from "Numerical Methods in C" ***

    [ma,~]=size(param.a);
    oneda=zeros(ma);
    covar=zeros(ma,ma);
    da=zeros(ma);

    %initialize on first pass based on alamda setting

    if (param.alamda <0.0)
        %clear all working matrices
        param.mfit=0;
        for i=1:ma
            if (param.ia(i)>0)
                param.mfit=param.mfit+1;
            end
        end
        param.alamda=0.001;

        param.alpha=zeros(ma,ma);
        param.beta=zeros(ma);

        %fun first pass
        [param.alpha,param.beta,param.chisq]=...
            mrqcof(param,param.a,param.alpha,param.beta,func);
```

```
      %save computed cost for next run
      param.ochisq=param.chisq;
      param.atry=param.a;
end

temp=zeros(param.mfit,param.mfit);
for j=1:param.mfit
    for k=1:param.mfit
        covar(j,k)=param.alpha(j,k);
    end
    covar(j,j)=param.alpha(j,j)*(1+param.alamda);
    for k=1:param.mfit
        temp(j,k)=covar(j,k);
    end
    oneda(j)=param.beta(j);
end

%get the solution matrix
[temp,oneda]=gaussj(temp,oneda);
for j=1:param.mfit
    for k=1:param.mfit
        covar(j,k)=temp(j,k);
    end
    da(j)=oneda(j);
end

%Once converged evaluate convariance matrix
if (param.alamda==0)
    covar=covsrt(covar,param.ia,paramm.mfit);
    param.alpha=covsrt(param.alpha,param.ia,param.mfit);
else
    j=0;
    for l=1:ma
        if (param.ia(l)>0)
            param.atry(l)=param.a(l)+da(j+1);
            j=j+1;
        end
    end
```

```
        [covar,da,param.chisq]=mrqcof(param,param.atry,covar,da,func);
    if (param.chisq<param.ochisq)
        %converging

        %tighten the spread
        param.alamda=param.alamda*0.1;
        param.ochisq=param.chisq;
        for j=1:param.mfit
            for k=1:param.mfit
                %save cost for next iteration
                param.alpha(j,k)=covar(j,k);
            end

            %save new values for next run
            param.beta(j)=da(j);
        end
        for l=1:ma
            param.a(l)=param.atry(l);
        end
    else
        %diverging

        %loosen up the spread
        param.alamda=param.alamda*10;

        %restore old error metrix
        param.chisq=param.ochisq;
    end
  end
end
```

D.15 NORMQ(·)

```
function qOut= NormQ(qIn)
%NormQ() - normalize a quaterion

mag=sqrt(qIn'*qIn);
for i=1:4
    qOut(i)=qIn(i)/mag;
end
```

D.16 ORTHOG(·)

```
function b = Orthog(a)
%Orthog() - orthoganolize a 3x3 matrix

ata=a'*a;
c=sqrt(3/(ata(1,1)+ata(2,2)+ata(3,3)))*a;
for i=1:4
    b=(3/2)*c-0.5*c*transpose(c)*c;
    c=b;
end
```

D.17 OUTERPRODUCT(·)

```
function c = OuterProduct(a,b)
%OuterProduct() - compute the outer product of two vectors
%the outer product is the nxn matrix product of a*transpose(b)

N=length(a);
c=zeros(N,N);
for i=1:N
    for j=1:N
        c(i,j)=a(i)*b(j);
    end
end
```

D.18 QMUL(·)

```
function q = Qmul(r,s,i,j)
%Quaternion Multiply
%Multiply two quaternions.  The i and j flags can be
%set to 1 to invert the associated quaternion input
%value before the multiplication
if (i>0)
    r(1)=-r(1);
end

if (j>0)
    s(1)=-s(1);
end
```

```
q(1,1)=r(1)*s(1)-r(2)*s(2)-r(3)*s(3)-r(4)*s(4);
q(2,1)=r(1)*s(2)+r(2)*s(1)+r(3)*s(4)-r(4)*s(3);
q(3,1)=r(1)*s(3)-r(2)*s(4)+r(3)*s(1)+r(4)*s(2);
q(4,1)=r(1)*s(4)+r(2)*s(3)-r(3)*s(2)+r(4)*s(1);
```

D.19 QTOA(·)

A function to convert quaternios to rotation matrices.

```
function A = QtoA(q)
%Quaternion to Aer (Euler Angles)
%convert a quaternion orientation to Euler Angles (in radians)

A(1,1)=q(1)*q(1)+q(2)*q(2)-q(3)*q(3)-q(4)*q(4);
A(1,2)=2*(q(2)*q(3)-q(1)*q(4));
A(1,3)=2*(q(2)*q(4)+q(1)*q(3));
A(2,1)=2*(q(1)*q(4)+q(2)*q(3));
A(2,2)=q(1)*q(1)-q(2)*q(2)+q(3)*q(3)-q(4)*q(4);
A(2,3)=2*(q(3)*q(4)-q(1)*q(2));
A(3,1)=2*(q(2)*q(4)-q(1)*q(3));
A(3,2)=2*(q(1)*q(2)+q(3)*q(4));
A(3,3)=q(1)*q(1)-q(2)*q(2)-q(3)*q(3)+q(4)*q(4);


qn=q(1)*q(1)+q(2)*q(2)+q(3)*q(3)+q(4)*q(4);


for i=1:3
    for j=1:3
        A(i,j)=A(i,j)/qn;
    end
end
```

D.20 QTOAER(·)

```
function aer = QtoAer(q)
%QtoAer - convert a quaternion to Euler angles (radians)

A=QtoA(q);
aer=AtoAer(A);
```

D.21 RFINDR(·)

Find the position given a field measurement and the interpolation cube parameters.

```
function rs = rFindR(rs,svm,xyz,vals)
%rFindR()   An LMM function to solve for a position inside the
%           interpolation volume using the enhanced det/trace/sts method
%rs         seed
%xy         positions of cube vertices
%vals       field vectors at cube vertices

mVal=GetBtB(svm)';

[ma,~]=size(rs);
[npt,~]=size(mVal);

ia=ones(ma,1);
sig=ones(npt,1);

param = CreateLMM_ParamStruct(rs,mVal,ia,sig,0.001,2,100);
cFunc=@rFindr_Cost;
param=RunLMM(param,cFunc);
rs=param.a;

    function    eVal=rFindr_CostUtil(r)
        %this function is assembles the various values used
        %for the cost function analysis
        sve=Interp(r,xyz,vals)';
%       s4e=[sve(1:3)';sve(4:6)';sve(7:9)'];
%       eVal=[det(s4e);svd(s4e);GetBtB(sve)'];
        eVal=GetBtB(sve)';
    end

    function [ymod,dyda] = rFindr_Cost(~,r)
        %this function provides a cost function for the LMM solver test

        ymod=rFindr_CostUtil(r);
        delta=10e-6;
        dyda=zeros(npt,ma);
        for i=1:ma
            rtry=r;
```

```
                rtry(i)=rtry(i)+delta;
                dyda(:,i)=rFindr_CostUtil(rtry);
                dyda(:,i)=(dyda(:,i)-ymod)/delta;
        end
    end
end
```

D.22 RUNLMM(·)

Solve an equation with given parameters using Minpac.

```
function param = RunLMM( param,func )
%This is the top level function that calls LMM to minimize the error in
% a non-linear multivariant optimization


%initialize wieghting coef
%the standard deviation is used as a wieghting coef when computing the
%chisq funciton. You could modify the wieghting coef based on some
%criteria to improve convergence on an estimate


%initialize mrqmin function
param.alamda=-1;                    %set up for initialization
param=mrqmin(param,func);                  %execute first iteration
param.iter=1;
itst=1;


flag=1;
while(flag==1)
    param.iter=param.iter+1;    %increment the iteration count

    %save previous error value
    ochisq=param.chisq;

    %get next estimate
    param=mrqmin(param,func);                %next iteration

    %see if we have reduced the error from the previous iteration.  If the
    %error increased, clear the convergence count and try again.  If the
    %error was reduced, increment the convergence counter.
    if ((abs(ochisq-param.chisq)<param.ConvThresh)&&(ochisq ~=param.chisq))
        %this iteration resulted in an improvement less than the
```

```
    %convergence threshold so increment the convergence count
        itst=itst+1;
    else
        itst=0;
    end

    %check for convergence
    %convergence is declared when teh improvement falls below the
    %covergence threshold for consecutive iterations
    if ((itst<param.ConvLimit)&&(param.iter<=param.IterLimit))
        %continue looping
        flag=1;
    else
        %exit
        flag=0;
    end
end
```

D.23 SOLVEFORORIENTATION(·)

```
function qs = SolveForOrientation(sest,smeas)
%SolveForOrientation()- find the ideal rotation for a position estimate
%this function is part of the LMM mapping PnO solver and finds the
%quaternion orientation that gives the best fit of the unrotated estimated
%S4 to the measured S4.

%estimate the orientation to estimate the seed
Aseed=
(smeas/sest)';
Aseed=Orthog(Aseed);
qSeed=AerToQ(AtoAer(Aseed));

%generate the vectorized estimate
svec=[sest(1,:)';sest(2,:)';sest(3,:)'];

%initialize and launch the optimization
[ma,~]=size(qSeed);
[npt,~]=size(svec);
ia=ones(ma,1);
sig=ones(npt,1);
```

```matlab
param = CreateLMM_ParamStruct(qSeed,svec,ia,sig,0.001,2,100);
cFunc=@SolveForQ_cost;
param=RunLMM(param,cFunc);
qs=param.a;
qs=qs/sqrt(qs'*qs);

    function y=SolveForQ_util(q)
        q=q/sqrt(q'*q);
        se=(AerToA(QtoAer(q))'*sest);
        y=[se(1,:);se(2,:);se(3,:)];
    end

    function [ymod,dyda]=SolveForQ_cost(~,qe)
        %get the ideal rotation at the estimated position
        yn=SolveForQ_util(qe);
        ymod=[yn(1,:)';yn(2,:)';yn(3,:)'];
        dyda=zeros(npt,ma);
        delta=10e-6;
        for i=1:ma
            qtry=qe;
            qtry(i)=qtry(i)+delta;
            yn=SolveForQ_util(qtry);
            dyda(:,i)=[yn(1,:)';yn(2,:)';yn(3,:)'];
            dyda(:,i)=(dyda(:,i)-ymod)/delta;
        end
    end
end
```

D.24 SOLVESTS(·)

```matlab
function x = SolveStS(r,MapStruct,smeas)

%breakout the measured S4 and generate the StS
svm=[smeas(1,:)';smeas(2,:)';smeas(3,:)'];
%mVal=[det(smeas);svd(smeas);GetBtB(svm)'];
mVal=GetBtB(svm)';

[ma,~]=size(r);
[npt,~]=size(mVal);
```

```
ia=ones(ma,1);
sig=ones(npt,1);


param = CreateLMM_ParamStruct(r,mVal,ia,sig,0.001,2,100);
cFunc=@SolveStS_Cost;
param2=RunLMM(param,cFunc);
x=param2.a;



    function   eVal=SolveStS_CostUtil(r)
        %this function is assembles the various values used
        %for the cost function analysis
        sve=GetField(r,MapStruct);
%        s4e=[sve(1:3)';sve(4:6)';sve(7:9)'];
%        eVal=[det(s4e);svd(s4e);GetBtB(sve)'];
        eVal=GetBtB(sve)';
    end

    function [ymod,dyda] = SolveStS_Cost(~,r)
        %this function provides a cost function for the LMM solver test
        ymod=SolveStS_CostUtil(r);
        delta=10e-6;
        dyda=zeros(npt,ma);
        for i=1:ma
            rtry=r;
            rtry(i)=rtry(i)+delta;
            dyda(:,i)=SolveStS_CostUtil(rtry);
            dyda(:,i)=(dyda(:,i)-ymod)/delta;
        end
    end
end
```

References

[1] A. Kiruluta, M. Eizenman, and S. Pasupathy, "Predictive head movement tracking using a Kalman filter," *IEEE Transactions on Systems, Man and Cybernetcs – Part B: Cybernetics,* vol. 27, no. 2 Apr. 1997, pp. 326–331. DOI: 10.1109/3477.558841. 1, 2

[2] R.H.Y. So,and M.J. Griffin, "Experimental studies of the use of phase lead filters to compensate lags in head-soupled visual displays," *IEEE Transactions on Systems, Man and Cybernetics,* Part A. vol. 26, no. 4, pp. 445–454, Jul 1996. DOI: 10.1109/3468.508823. 1, 3

[3] J. Bohg, *Real-times structure from motion using Kalman filtering,* Ph.D. Dissertation, 2005. 1, 2

[4] J. S. Goddard Jr. *Pose and motion estimation from vision using dual quaternion-based extended Kalman filtering,* Ph.D. Dissertation, 1997. 2, 6, 51, 67

[5] W. C. Chang and C. W. Cho, "Active head tracking using integrated contour and template matching in indoor cluttered environment," *IEEE Conference on Systems, Man and Cybernetics,* vol. 6, pp. 5167–5172, Oct. 11, 2006. DOI: 10.1109/ICSMC.2006.385128. 1, 2

[6] R. F. Stengel, *Stochastic optimal control,* Wiley and Sons, 1986. 4

[7] A. V. Rhijn, R. V. Liere, and J. D. Mulder, "An analysis of orientation prediction and filtering methods for VR/AR," *Proceedings of the IEEE Virtual Reality Conference,* 2005, pp. 67–74. DOI: 10.1109/VR.2005.1492755. 1, 27

[8] J. R. Wu and M. Ouhyoung, "On latency compensation and its effects on head-motion trajectories in virtual environments," *The Visual Computer,* vol. 16, pp. 79–90, 2000. DOI: 10.1007/s003710050198. 1, 25

[9] R. H. Y. So, "Lag compensation by image deflection and prediction: a review on the potential benefits to virtual training applications for manufacturing industry," Hong Kong University of Science and Technology, Hong Kong, unpublished.

[10] R. H. Y. So, W. T. Lo, and A. T.K. Ho, "Effects of navigational speed on motion sickness caused by an immersive virtual environment," *Human Factors,* Fall 2001, vol. 43, no. 3, pp. 452–461. DOI: 10.1518/001872001775898223. 1, 25

[11] H. Himberg, Y. Motai, and C. Barrios, "R-adaptive Kalman filtering approach to etimate head orientation for driving simulator," *IEEE Intelligent Transportation Systems Conference,* Sept. 2006, pp. 851–857. DOI: 10.1109/ITSC.2006.1706850. 2, 26

[12] Y. Zhang, H. Hu, and H. Zhou, "Study on adaptive Kalman filtering and algorithms in human movement tracking," *Proceedings of the IEEE International Conference on Information Acquisition,* 2005. DOI: 10.1109/ICIA.2005.1635045. 1, 3

[13] G. Welch and G. Bishop, "An introduction to the Kalman filter," *SIGGRAPH,* 2001, Course notes. 4, 28

[14] R. S. Allison, L. R. Harris, M. Jenkin, U. Jasiobedzka, and J. E. Zacher, "Tolerance of temporal delay in virtual environments," *IEEE Virtual Reality Conference,* 2001, pp. 247. DOI: 10.1109/VR.2001.913793. 1

[15] J. Y. Jung, B. D. Adelstein, and S. R. Ellis, "Discriminability of prediction artifacts in a time-delayed virtual environment," *Proceedings of IEA,* 2000/HFES 2000, pp. 499–502. DOI: 10.1177/154193120004400504. 1, 25

[16] J. C. K. Chou, "Quaternion kinematic and dynamic differential equations" *IEEE Transactions on Robotics and Automation,* vol. 8, no. 1, pp. 53–64, February 1992. DOI: 10.1109/70.127239. 2, 6, 27, 28, 51

[17] E. A. Coutsias and L. Romero, "The quaternions with an application to rigid body dynamics," Lecture notes for seminar on dynamic and stability theory, Dept. of Mathmatics, Univ. of New Mexico, February 1999 (Sandia Technical Report, SAND2004–0153, 2004). 4

[18] A. L. Schwab, "Quaternions, finite rotation and euler parameters," 2002, unpublished.

[19] M. D. Shuster, "A survey of attitude representations," *The Journal of the Astronautical Sciences,* vol. 41, no. 4, October-December 1993, pp. 439–517. 4

[20] Y. Bar-Sholon, X. R. Li, and T. Kirubarajan, *Estimation with Applications to Tracking and navigation,* Wiley and Sons, 2001. DOI: 10.1002/0471221279. 4, 29

[21] Y.-J. Cheon and J.-H. Kin, "Unscented filtering in a unit quaternon space for spacecraft attutude estimateion," *IEEE Internation Symposium on Industrial Electronics,* June 2007, pp. 66–71. DOI: 10.1109/ISIE.2007.4374575. 3, 46

[22] R. Azuma and F. Bishop, "A frequency-domain analysis of head-motion prediction," *NSF/ARPA Science and Technolog Center for Computer Graphics and Visulalization,* (ARPA contrat DABT63–93–C-C048). DOI: 10.1145/218380.218496. 1

[23] R. S. Allison, L. R. Harris, M. Jenkin, U. Jasiobedzka and J. E. Zacher, "Tolerance of temporal delay in virtual environments," *Proc. IEEE Int. Conf. on Virtual Reality*, (2001), pp. 247–254. DOI: 10.1109/VR.2001.913793. 25

[24] P. M. Jaekl, R. S. Allison, L. R. Harris, U.T . Jasiobedzka, H. Jenkin, M. R. Jenkin, J. E. Zacher and D. C. Zikovitz, "Perceptual stability during head movement in virtual reality," *IEEE Int.Conference on Virtual Reality*, 4 (2002), pp. 149–155. DOI: 10.1109/VR.2002.996517. 1

[25] A. Ude, "Filtering in a unit quaternion space for model-based object tracking," *Robotics and Autonomous Systems*, vol. 28, no. 2–3, pp. 163–172, August 1999. DOI: 10.1016/S0921-8890(99)00014-7. 1, 4

[26] J. LaViola, "A comparison of unscented and extended Kalman filtering for estimating quaternion motion." *Proceedings of the 2003 American Control Conference*, (Jun 2003), IEEE Press, pp. 2435–2440. DOI: 10.1109/ACC.2003.1243440. 1, 2

[27] F. Ababsa, M. Mallem, and D. Roussel, "Comparison Between Particle Filter Approach and Kalman Filter-Based Technique For Head Tracking in Augmented Reality Systems." In *Proceedings of the 2004 IEEE International Conference on Robotics and Automation (ICRA2004)*. USA, May 2004, 1021—1026. DOI: 10.1109/ROBOT.2004.1307284. 2

[28] J. Liang, C. Shaw and M. Green, "On temporal-spatial realism I the virtual reality environment," *Proceedings of 4th Annual Symposium on User Interface Software and Technology*, Nov. 1991, pp. 19–25. DOI: 10.1145/120782.120784. 1, 2, 25, 26, 52

[29] S. B. Choe and J. J. Faraway, "Modeling head and hand orientation during motion using quaternions," *SAE Transactions*, Journal of Aerospace, 2004–01-2179. DOI: 10.4271/2004-01-2179. 4

[30] R. Azuma and G. Bishop, "Improving static and dynamic registration in a see-through hmd," *Proceedings of SIGGRAPH*, 1994, pp. 197–204. DOI: 10.1145/192161.192199. 1, 2

[31] P. S. Maybeck, *Stochastic models, estimation and control*, Academic Press, 1979. 4, 28

[32] K. Terada, A. Oba, and A. Ito, "3D human head tracking using hypothesized polygon model," *IEEE Conference on Systems, Man and Cybernetics*, 2005. vol. 2, pp. 1396–1401, Oct. 2005. DOI: 10.1109/ICSMC.2005.1571342. 1

[33] J. -R. Wu and M. Ouhyoung, "A 3D tracking experiment on latency and its compensation methods in virtual environments," *Proceedings of the 8th annual ACM symposium on User interface and software technology*, Nov. 1994, pp. 41–49. DOI: 10.1145/215585.215650. 1, 2

[34] J. LaViola, "Double exponential smoothing: An alternative to Kalman filter-based predictive tracking," *ACM International Conference Proceeding Series,* vol. 39, 2003, pp. 199–206. DOI: 10.1145/769953.769976. 3

[35] J. Marins, X. Yn, E. Bachmann et al., "An extended Kalman filter for quaternions-based orientation using MARG sensors," *Proceeedings of the International Conference on Intelligent Robots and Systems,* vol. 4, Nov. 2001, pp. 2003–2011. DOI: 10.1109/IROS.2001.976367. 3, 27

[36] A. Sabatini, "Quaternion-based extended Kalman filter for determining orientation by inertial and magnetic sensing," *IEEE Transactions on Biomedical Engineering,* vol. 53, no. 7, July 2006, pp. 1346–1356. DOI: 10.1109/TBME.2006.875664. 3, 27

[37] K. Ali, C. Vanelli, J. Biesiadecki, M. Maimone, U. Y. Cheng, A. Martin and J. Alexander, "Attitude and position estimation on the mars exploration rovers," *IEEE International Conference on Systems, Man and Cybernetics,* vol.1, Oct. 2005, pp. 20–27. DOI: 10.1109/ICSMC.2005.1571116. 3

[38] E.J. Lefferts, F.L. Markley, M.D. Shuster, "Kalman Filtering for Spacecraft Attitude Estimation, " *J. Guidance, Control, and Dynamics,* vol. 5, pp. 417–429, Sep.-Oct. 1982, DOI: 10.2514/3.56190. 26

[39] I.Y. Bar-Itzhack, Y. Oshman, "Attitude Determination from Vector Observations: Quaternion Estimation, " *IEEE Trans. Aerospace and Electronic Systems,* vol.AES-21, pp. 128–136, Jan.1985. DOI: 10.1109/TAES.1985.310546. 26, 27

[40] G. Burdea, "Invited review: The synergy between virtual reality and robotics," *IEEE Trans. Robotics and Automation,* vol. 15, no. 3, pp. 400–410, Jume 1999,. DOI: 10.1109/70.768174.

[41] V. P. Jilkov, X. R. Li, "Bayesian estimation of transition probabilities for markovian jump systems by stochastic simulation," *IEEE Trans. Signal Processing,* vol. 52, issue 6, pp. 1620–1630, June 2004. DOI: 10.1109/TSP.2004.827145. 30

[42] V.P. Jilkov, X. R. Li, "Adaptation of transistion probability matrix for multiple model estimators," *Proc. Inter. Conf. Information Fusion,*

[43] C. Nielson, M. Goodrich, R. Ricks, "Ecological interfaces for improving mobile robot teleoperation," *IEEE Trans. Robotocs,* vol. 23, no. 5, pp. 927–941, 2007. DOI: 10.1109/TRO.2007.907479.

[44] M. Shuster, "A survey of attitude representations," *J. Astronautical Sciences,* vol. 41, no. 4, pp. 439–517, Oct.-Dec. 1993. 27

[45] D. Kyger, P. Maybeck, "Reducing lag in virtual displays using multiple model adaptive estimation," *IEEE Trans. Aerospace and Electronic Systems*, vol. 34, no, 4. Oct. 1998. DOI: 10.1109/7.722711. 25, 26, 35, 42

[46] L. C. Yang, J. H. Yang, E. M. Feron, "Multiple Model Estimation for Improving Conflict Detection algorithms," *Proc. IEEE Conf. Systems, Man and Cybernetecis*, vol. 1, pp. 242–249, Oct. 2004. DOI: 10.1109/ICSMC.2004.1398304. 27

[47] H. Himberg, Y. Motai, "Head orientation prediction: delta quaternions versus quaternions," *IEEE Trans. Systems, Man and Cybernetecis Part B*, In press. DOI: 10.1109/TSMCB.2009.2016571. 26, 29, 51

[48] C. Hilde, T. Moore, M. Smith, "Multiple Model Kalman Filtering for GPS and Low-cost INS integration," Institute of Engineering, Surveying and Space Geodesy, University of Nottingham, 2004. 25, 32

[49] K. Ali, C. Vanelli, J. Biesiadecki, M. Maimone, U. Y. Cheng, A. Martin and J. Alexander, "Attitude and position estimation on the mars exploration rovers," *Proc. IEEE Conf. Systems, Man and Cybernetics*, vol.1, pp. 20–27, Oct. 2005. 27

[50] E. Derbez, B. Remillard, A. Jouan, "A comparison of fixed gain IMM against two other filters," *Proc. IEEE Int. Conf. Information Fusion*, vol. 1, July 2000. DOI: 10.1109/IFIC.2000.859872. 29

[51] X. R. Li, V. P. Jilkov, "A Survey of Maneuvering Target Tracking – Part V: Multiple Model Methods," *IEEE Trans. Aerospace and Electronic Systems*, vol. 41, issue 4, pp. 1255–1321, Oct 2005. DOI: 10.1109/TAES.2005.1561886. 25, 30

[52] E. Daeipour, Y. Bar-Shalom, "Adaptive beam pointing of phased array radar using IMM estimator," *Proc. American Control Conf.*, Baltimore, MD, 1994. DOI: 10.1109/ACC.1994.752445. 33

[53] H. Bom, Y. Bar-Shalom, "The interacting multiple model algorithm for systems with Markovian switching coefficients," *IEEE Trans. Automatic Control*, vol. 33, issue 8, pp. 780–783, Aug. 1988. DOI: 10.1109/9.1299. 25, 29

[54] L. Campo, P. Mookerjee, Y. Bar-Shalom, "State estimation for systems with sojourn-time-dependent markov model switching," *IEEE trans. Automatic Control*, vol. 36, no. 2, pp. 238–243, Feb. 1991. DOI: 10.1109/9.67304. 30

[55] W. D. Blair, G. A. Watson, "IMM algorithm for solution to benchmark problem for tracking maneuvering targets," *Proc. Acquisition, Tracking, and Pointing IX, SPIE 2221*, Orlando, Florida, pp. 476–488, 1994. 29

[56] E. Mazor, A. Averbuch, Y. Bar-Shalom, J. Dayan, "Interacting multiple model methods in target tracking: a survey," *IEEE Trans. Aerospace and Electronic Systems*, vol. 34, no. 1, pp. 103–123, Jan. 1998. DOI: 10.1109/7.640267. 29

[57] X. R. Li, Y. Zhang, "Numerically robust implementation of multiple-model algorithms," *IEEE Trans. Aerospace and Electronic Systems*, vol. 36, no. 1, pp. 266–278, January 2000. DOI: 10.1109/7.826329. 31

[58] J. Wen, K. Kreutz-Delgado, "The attitude control problem," *IEEE Trans.Automatic Control*, vol. 36, no. 10, pp. 1148–1162, Oct. 1991. DOI: 10.1109/9.90228. 27

[59] Y. Cheon, J. Kin, "Unscented filtering in a unit quaternion space for spacecraft attutude estimation," *IEEE Int. Symp. Industrial Elect.*, pp. 66–71, June 2007. DOI: 10.1109/ISIE.2007.4374575. 27

[60] R. Thompson, I. Reid, L. Munoz, D. Murray, "Providing Synthetic views for teleoperation using visual pose tracking in multiple cameras," *IEEE Trans. Systems, Man, and Cybernetics – Part A: Systems and Humans*, vol. 31, no. 1, pp. 43–54, Jan. 2001. DOI: 10.1109/3468.903865.

[61] J. Kofman, X. Wu, T. Luu, S. Verma, "Teleoperation of a robot manipulator using vision-based human-robot interface," *IEEE Trans. Industrial Electronics*, vol. 52, no. 5, pp. 1206–1219, Oct. 2005. DOI: 10.1109/TIE.2005.855696.

[62] F.L. Markley,, "Attitude error representations for Kalman filtering," *J. Guidance, Control, and Dynamics*, pp. 311–317, vol. 26, Mar.-Ap. 2003 27

[63] M.E. Pittelkau, "An analysis of the quaternion attitude determination filter," *J. Astronautical Sciences*, vol. 51, issue 1, pp. 103–120, Jan.-Mar., 2003. 27

[64] T. Kirubarajan, Y. Bar-Shalom, "Kalman filter versis IMM estimator: when do we need the later?," *IEEE Trans. Aerospace and Electronic Systems*, vol. 39, no. 4, pp. 1452–1457, Oct. 2003 DOI: 10.1109/TAES.2003.1261143. 37

[65] D. Frantz, A. Wiles, S. Leis, S. Kirsch, "Accuracy assessment protocols for electromagnetic tracking systems," *Physics in Medicine and Biology*, July 2003, vol. 48, issue 14, pp. 2241–2251. DOI: 10.1088/0031-9155/48/14/314. 50

[66] N. Hagemeister, G. Parent, S. Husse, et al., "A simple and rapid method for electromagnetic field distortion correction when using two Fastrak sensors for biomechanical studies," *Journal of Biomechanics*, 2008, vol. 41, issue 8, pp. 1813–1817. DOI: 10.1016/j.jbiomech.2008.02.030. 51

[67] M. Nixon, B. McCallum, W. Fright, et al., "The effects of metals and interfering fields and on electromagnetic trackers," *Presence: Tele-Operators and Virtual Environments*, Apr. 1998, vol. 7, issue 2, pp. 204–218. DOI: 10.1162/105474698565587. 47

[68] X. Wu, R.Taylor, "A direction space interpolation technique for calibration of electromagnetic surgical navigation systems," *MICCAI 2003*, part 2, Lecture notes on Computer Science, vol. 2879, pp. 215–222. DOI: 10.1007/978-3-540-39903-2_27. 51

[69] J. Day, D. Murdoch, G. Dumas, "Calibration of position and angular data from a magnetic tracking device," *Journal of Biomechanics*, August 2000, vol. 33, issue 8, pp. 1039–1045. DOI: 10.1016/S0021-9290(00)00044-0. 50

[70] H. Wang, G. Jiang, "Study on sensor array applied in electromagnetic tracking system," *IEEE/ICME International Conference on Complex Medical Engineering*, 2007, pp. 180–192. DOI: 10.1109/ICCME.2007.4381719. 51

[71] M. Ghazisaedy, D. Adamczyk, D. Sandin, R. Kenyon, T. DeFanti, "Ultrasonic calibration of a magnetic tracker in a virtual reality space," *IEEE Virtual Reality Annual International Symposium*, March 1995, pp. 179–188. DOI: 10.1109/VRAIS.1995.512494. 50

[72] M. Ikits, J.D. Brederson, C.D. Hansen, J.M. Hollerbach,"An Improved Calibration Framework for Electromagnetic Tracking Devices," *IEEE Proceedings on Virtual Reality*, March 2001, pp. 63–70. DOI: 10.1109/VR.2001.913771. 49

[73] H. Jones, "Method and apparatus for determining remote object orientation and position," US Patent 4,737,794, Apr. 1988. 52

[74] F. Raab, E. Blood, T. Steiner, H. Jones, "Magnetic position and orientation tracking system," *IEEE Transactions on Aerospace and Electronic Systems*, Sept. 1979, vol. AES-15, no. 5, pp. 709–718, Sept. 1979 DOI: 10.1109/TAES.1979.308860. 49

[75] W. Birkfellner, F. Watzinger, F. Wanschitz, et al.," Systematic distortions in magnetic position digitizers," *Medical Physics*, Nov. 1998, vol. 25, issue: 11, pp. 2242–2248. 48

[76] W. Birkfellner, F. Watzinger, F. Wanschitz, et al.,"Calibration of tracking systems in a surgical environment," *IEEE Transactions on Medical Imaging*, Oct 1998, vol. 17, issue: 5, pp. 737–742. DOI: 10.1109/42.736028. 49

[77] J. Hummel, M. Figl, C. Kollmann, et al., "Evaluation of a minature electromagnetic position tracker," *Medical Physics*, Oct 2002, vol. 29, issue 10, pp. 2205–2212. DOI: 10.1118/1.1508377. 48

[78] J. Hummel, M. Bax, M.Figl, et al. "Design and application of an assessment protocol for electromagnetic systems," *Medical Physics*, Jul 2005, vol. 32, issue 7, pp. 2371–2379. DOI: 10.1118/1.1944327.

[79] U. Jayaram, R. Repp, "Integrated real-time calibration of electromagnetic tracking of user motions for engineering applications in virtual environments," *Journal of Mechanical Design*, Dec 2002, vol. 124, issue 4, pp. 623–632. DOI: 10.1115/1.1517562. 50

[80] M. Livingston, A. State, "Magnetic tracker calibration for improved augmented reality registration," *Presence: Teleoperators and Virtual Environments*, Oct 1997, vol. 5, issue 5, pp. 532–546. 50

[81] M. Feuerstein, T. Reichl, J. Vogel, et al. "Magneto-optical tracking of flexible laparoscopic ultrasound: model-based detection and correction of magnetic tracking errors," *IEEE Transactions on Medical Imaging*, Jun 2009, vol. 28, issue 6, pp. 951–967. DOI: 10.1109/TMI.2008.2008954. 50

[82] A. Milne, D. Chess, J. Johnson et al., "Accuracy of an electromagnetic tracking device: a study of the optimal operating range and metal interference," *Journal of Biomechanics*, June 1996, vol. 29, issue 6, pp. 791–793. DOI: 10.1016/0021-9290(96)83335-5. 48

[83] K. Nakada, M. Nakamoto, Y. Sato, et al., "A rapid method for magnetic tracker calibration using a magneto-optic tracker," *Medical Image Computing and Computer-Assisted Intervention: Lecture notes in Computer Science*, 2003, vol. 2879, pp. 285–293. DOI: 10.1007/978-3-540-39903-2_36. 49

[84] A. Wagner, K. Schicho, W. Birkfellner, et al., " Quantitative analysis of factors affecting intraoperative precision and stability of optoelectronic and electromagnetic tracking systems," *Medical Physics*, May 2002, vol. 29, issue 5, pp. 905–912. DOI: 10.1118/1.1469625. 48

[85] V. Kindratenko, "A survey of electromagnetic position tracker calibration techniques," *Virtual Reality: Research, Development and Application*, 2000, vol. 5, no. 3, pp. 169–182. DOI: 10.1007/BF01409422. 49

[86] V. Ochoa-Mayorga, P. Boulanger, M. Garcia, "Local quaternion weighted difference funcions for orientation calibration on elctromagnetic trackers," *IEEE International Workshop on Computational Advances in Multi-Sensor Adaptive Processing*, 2005, pp. 233–236. DOI: 10.1109/CAMAP.2005.1574227. 50

[87] H. Jones, R. Higgins, H. Himberg, "Magnetic position and orientation measurement system with eddy current distortion compensation," US Patent 7,292,948 B2, Nov. 2006 49

[88] Franke R, Nielson G., "Smooth interpolation of large sets of scattered data," *Internatinal Journal for Numberical Methods in Engineering*, vol. 15, issue 11, pp. 1691–1704, 1980. DOI: 10.1002/nme.1620151110.

[89] F.S. Grant, G.F. West, *Interpertation theory in applied geophysics*, McGraw-Hill, 1965.

[90] H. R. Jones, I. Khalfin, "Methods and apparatus for electromagnetic position and orientation tracking with distortion compensation," US Patent 6,369,564 B1, Apr. 2002. 49

Authors' Biographies

HENRY HIMBERG

Henry Himberg received the B.S. degree in electrical and computer engineering from Clarkson University, Potsdam, NY, in 1982, the M.S. degree from the University of Vermont, Burlington, in 2005, and the Ph.D. degree in electrical and computer engineering from Virginia Commonwealth University, Richmond, in 2010. He is currently a Staff Scientist with Polhemus, Colchester, VT. His research interests include signal processing, position/orientation measurement, and motion prediction.

YUICHI MOTAI

Yuichi Motai received the B.Eng. degree in instrumentation engineering from Keio University, Tokyo, Japan, in 1991, the M.Eng. degree in applied systems science from Kyoto University, Kyoto, Japan, in 1993, and the Ph.D. degree in electrical and computer engineering from Purdue University, West Lafayette, in 2002. He is currently an Associate Professor of Electrical and Computer Engineering at Virginia Commonwealth University, Richmond, VA. His research interests include the broad area of sensory intelligence, particularly in medical imaging, pattern recognition, computer vision, and sensory-based robotics.